Thermal Characteristics of the Moon

**Progress in
Astronautics and Aeronautics**

Martin Summerfield,
Series Editor
PRINCETON UNIVERSITY

(Other volumes are planned.)

The MIT Press
Cambridge, Massachusetts,
and London, England

Progress in
Astronautics and Aeronautics

An American Institute of Aeronautics
and Astronautics Series

Martin Summerfield, Series Editor

Volume 28

Thermal Characteristics of the Moon

Edited by
John W. Lucas
JET PROPULSION LABORATORY

This book was printed
by The Alpine Press, Inc.
and bound by The Colonial Press, Inc.
in the United States of America.

Library of Congress Cataloging in Publication Data
Main entry under title:

Thermal characteristics of the moon.

(Progress in astronautics and aeronautics, v. 28)
Includes bibliographies.
1. Moon—Temperature and radiation. I. Lucas,
John W., 1923-   ed.   II. Series
TL507.P75 vol. 28 [QB588]   523.3'2   79-39803
ISBN 0-262-12058-5

Preface

Our knowledge of the moon has increased greatly during the past few years. The driving force behind this increase has been the space program. During the middle part of the decade of the 1960's the United States Ranger spacecraft series produced a quantum jump in our knowledge; the later Surveyor spacecraft landings on the moon and the series of Lunar Orbiter spacecraft added larger quantum jumps. Two immediate uses for the information obtained from these spacecraft projects were to increase our scientific knowledge and to generate engineering information for utilization in spacecraft design and in operational planning of subsequent missions. The 1969–1971 Apollo flights resulted in still greater contributions to lunar knowledge, and these recently reached a peak with the very promising Apollo 15 landing. The USSR space effort also has produced major contributions to lunar knowledge, especially from the still-continuing Luna and Zond spacecraft series and from the Lunokhod rover that was first landed in 1970. Russian contributions included the first photographs of the far side of the moon, the return of a lunar soil sample, soil analyses with the rover, and microwave measurements of the lunar surface.

During these periods many techniques were employed to study the moon. Earth-based methods of oberving the moon were greatly extended and refined, spacecraft-borne experiments made direct measurements on the lunar surface, and simulated and actual lunar materials were subjected to tests in terrestrial laboratories. Finally, stimulated by all of these findings, theories of the origin of the moon and of its surface features were developed.

The purpose of this volume is to bring together in one place results obtained during this decade that bear on the thermal characteristics of the moon. The successive chapters show the way in which many pieces of information can be put together and used by different investigators from different disciplines to provide a coherent and reasonably accurate explanation of the moon's thermal characteristics. The characteristics include thermal properties of lunar surface material, surface and interior termperatures, heat flow patterns, and related physical properties. It is hoped that assembling the varied information together in one place in this volume will encourage further multidisciplinary studies and thus have a synergistic effect on future research.

For the purpose of this volume, leading investigators of lunar characteristics prepared up-to-date survey articles covering their own work and the works of others in their respective disciplines. Detailed information in each of the subject areas can be found through use of the reference lists provided in each chapter.

Section 1 contains chapters describing results of Earth-based measurements of the moon. Shorthill deals with the infrared portion of the lunar emission spectrum and describes the discovery of so-called "hot spots" on the lunar surface. Muhleman describes emission from the moon in the radio band and the relation to temperatures just below the lunar surface. Thompson and Zisk describe radar mapping of the moon and use this information further to describe the lunar hot spots.

Section 2 deals with in situ thermal measurements. Stimpson and Lucas present results obtained from analysis of telemetered engineering temperatures from Surveyor spacecraft; although these data provide 1000 times better resolution than Earth-based data, the derived lunar surface thermophysical properties were found to agree with Earth-based predictions. The Early Apollo Science Experiments Package placed on the lunar surface by the Apollo 11 astronauts contained an engineering experiment designated DTREM I, which was conceived as a spatial and temporal extension of Surveyor measurements on the lunar surface; Hickson describes this experiment and the lunar thermal data derived from it. The design and calibration of the Heat Flow Experiment of the Apollo Lunar Scientific Experiments Package (ALSEP) is given by Langseth, Drake, Nathanson, and Fountain. The Heat Flow Experiment was put in place on the lunar surface by the Apollo 15 astronauts in July 1971, and transmission of data of the direct measurement of the thermal gradient in the lunar surface began at that time.

As this volume went to press, the latest results of the Apollo 15 lunar Heat Flow Experiment described in Chapter 2c became available, and a brief report was prepared by Langseth et al. for insertion in the volume (Chapter 2d). The experiment was quite successful, even though as yet only one of the two probes driven into the lunar surface has produced the desired measurements. Analysis of the transmitted data received up to press time has yielded the result that the outward heat flux is not only larger than that calculated for an initially hot inert moon (cooling off by conduction and radiation) but also a good deal larger than that calculated for a moon composed of chondrites having typical concentrations of heat-producing radioactive isotopes. If this high level of outward heat flow and the corresponding high level of internal heat generation should be verified in future Apollo flights, it will indicate to planetary physicists that the moon's structure, and therefore its origin, may be more complex than hitherto suspected.

Section 3 is devoted primarily to studies of lunar-type materials in terrestrial laboratories. Wechsler, Glaser, and Fountain review the thermophysical properties of materials believed to simulate the lunar surface layer. Thermal conductivity data are emphasized, since there are large variations of this property in porous and particulate materials. It is pointed out that only after this information was obtained was it possible to discard the more extreme hypotheses of the nature of the lunar surface and to uphold the view that the lunar surface could support the pressure loads of man and his spacecraft. Thermophysical measurements on material brought from the moon by the Apollo 11 and 12 missions are summarized by Horai and Simmons. Results of these measurements are found to correspond in general to expected values. The rate of heat generation due to radioactivity is given and compared with terrestrial materials. Winter, Bastin, and Allen describe analyses of the effect of surface roughness and positive relief on lunar thermal characteristics. Observed behavior such as lunar hot spots, preferential rapid cooling of the lunar limb during an eclipse, and poleward darkening of the thermal microwave radiation from the moon are discussed in terms of several proposed models.

Section 4 consists of a single chapter by Reynolds, Fricker, and Summers in which theories of the thermal history of the moon are presented and discussed. Data from returned lunar materials, from the lunar magnetometer and seismometers, and from the Heat Flow Experiment should, in the near future, narrow down the number of possible thermal models of the moon.

Special thanks are due a number of individuals who provided invaluable help during the preparation of this volume. Dr. Martin Summerfield, Editor-in-Chief of the Progress in Astronautics and Aeronautics Series, originally suggested the theme for this volume and gave important guidance during its preparation. A. E. Wechsler and B. P. Jones, members of the AIAA Technical Committee on Thermophysics, which sponsored this volume, kindly gave overall technical advice. The Editorial Committee and the volume authors prepared reviews of the chapters. Miss Ruth F. Bryans, Director, Scientific Publications, provided assistance throughout the planning and preparation of this volume. My secretary, Miss Nancy Parmelee, continuously maintained up-to-date files and correspondence on all the chapters. A special note of appreciation is due my wife, Genevieve, and other members of my family whose patience and encouragement made the necessary effort possible.

John W. Lucas
November 1971

SECTION 1.   EARTH-BASED MEASUREMENTS

THE INFRARED MOON: A REVIEW

Richard W. Shorthill[*]

Boeing Scientific Research Laboratories, Seattle, Wash.

I. Introduction

The brightness temperature of the lunar surface has been determined using earth-based telescopes, balloon-borne telescopes, and aircraft-mounted telescopes. In June 1966, Surveyor I performed *in situ* thermal measurements followed by similar measurements at other locations on the moon with Surveyors III, V, VI, and VII in 1966 and 1968. With the two Apollo landings in 1969, and the one in 1970, we have passed from a remote sensing to a physical sampling era. Apollo 11, 12 and 14 provided samples of the lunar surface from which certain of its thermophysical properties have been determined.

Earth-based thermal observations revealed more than 1000 anomalous regions (hot spots) during an eclipse. It was also discovered that certain of the maria and portions of maria regions show an anomalous thermal behavior. In the coming Apollo and Luna missions, it can be expected that a landing will be made on or at least near one of the so-called hot spots and on one of the enhanced mare locations.

The purpose of this chapter is to describe the infrared measurements that have been made of the lunar surface. It is hoped that the material presented here and the accompanying references will be useful to various investigators in their interpretation of the thermophysical properties of the lunar surface.

The earliest measurement of infrared emission from the moon was made 100 years ago.[1] The first significant work was done between 1924 and 1928 by Pettit and Nicholson of the Mount Wilson Observatory[2]; their work has been reviewed in detail by

---

[*]Scientist, Environmental Sciences Laboratory.

others.[3-5] They reported measurements of brightness temperatures under various conditions: distribution of temperature along the lunar diameter at full moon, the subsolar point temperatures as a function of phase, the antisolar point (nighttime) temperature, and the transient temperature during a total lunar eclipse for a point near the disk center and at the limb. In the theoretical work that followed, Wesselink[6] and Jaeger and Harper[7,8] assumed that the thermophysical properties of the lunar surface were constant with depth and temperature. Even with a two-layer model[7] it was not possible to match simultaneously the lunation and eclipse measurements with one set of thermal parameters. These early theoretical studies, however, did suggest that the uppermost layer of the moon was of a porous or dust-like nature. Further measurements of eclipse cooling were made by Sinton and Strong in 1953,[9] and in 1958 and 1959 Sinton constructed isothermal contour maps over the lunar surface at nine different phases.[10] The lunar infrared measurements up to 1960, including a description of certain theoretical models, have been discussed by Sinton.[4]

During the lunar eclipse of March 13, 1960, Shorthill, Borough, and Conley[11] discovered that several rayed craters cooled less rapidly than their environs and, in particular, that Tycho was at least 40°K warmer than its environs an hour into the umbral phase. This surprising observation in the infrared, revealing differential properties over localized regions, provided an impetus to the author as well as several others to make additional measurements. Subsequently, extensive measurements were made during eclipses and during the lunar daytime. Measurements during the lunar night are less extensive. Some results of these measurements will be discussed, as well as a new thermophysical model developed by Winter and Saari[12] which fit both the lunation and eclipse data.

II. Experimental

Some of the problems and techniques associated with obtaining brightness temperatures of the lunar surface will be described in this section. Ground-based infrared measurements are complicated by absorption in the earth's atmosphere. A known temperature source outside the earth's atmosphere must be used for calibrating, or else all the loss factors between the lunar surface and the detector must be determined. The scanning techniques and position determination depend somewhat on the instrumentation.

A. Instruments

A thermopile was used to obtain the first lunar temperature
measurement.[1] Making reasonable assumptions of the moon's albedo and atmospheric losses, Rosse[1] in 1868 determined a
brightness temperature of 397°K. Pettit and Nicholson,[2] using
extremely fine thermocouples with a rock-salt window, determined a brightness temperature of the subsolar point to be
407°K. Using more modern infrared techniques, Sinton[13] developed a "pyrometer" for measuring planetary temperatures.
His system employs chopping rather than opposing junctions used
in the earlier measurements to eliminate drift. Figure 1 illustrates his instrument showing a tuning fork with reflecting
crystal filters C_1, C_2, C_3, and C_4 and transmission filter F.
Two Golay cells were used, G_1 and G_2, to look at two separated regions (the sky alone and the moon plus sky) in the focal
plane. As the chopper rotates, the images fall alternately on
G_1 and G_2, and only the difference in energy between the two
regions is amplified. A thermistor bolometer was used by
Shorthill to make the first measurements of the anomalous cooling of the ray crater Tycho.[5]

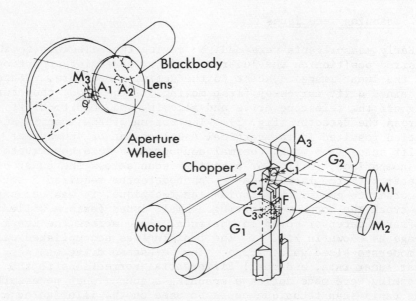

Fig. 1 Diagram of Sinton's pyrometer. A_1 and A_2 are the
two defining apertures.[13]

The thermocouple, Golay cell, and thermistor have a relatively slow time response, and measurements were thus limited
to a few features, or many hours were required to scan the entire disk. With the development of the germanium infrared
photodetectors of much greater sensitivity and speed, it became possible to design a system for acquiring measurements
over the disk in a reasonable length of time (15 to 30 min).
Figure 2 shows a cutaway view of the instrument developed by
Shorthill and Saari.[14,15] The incoming beam from the Newtonian
flat is interrupted by the polished aluminum chopper that re-
flects the radiation onto the aperture of the detector. Behind
the aperture is a series of f/5.6 baffles and a 10- to 12-μ
filter. A cone channel condenser finally transfers the radiation onto a mercury-doped germanium photodetector that was
cooled with liquid hydrogen or neon. The detector used for
the simultaneous measurement of the visible radiation was a
Dumont 6362 photomultiplier with the same field of view as the
infrared photodetector.

A more sensitive infrared bolometer using a gallium-doped
single-crystal germanium cooled to 2°K was developed by Low[16]
and is being used for lunar nighttime measurements by Low and
others.

B. Scanning Techniques

Early measurements were made by moving the telescope to the
desired position on the lunar surface by observing a portion
of the lunar image adjacent to the defining aperture. Sinton[10]
obtained drift curves by first moving off the moon; then turn-
ing off the telescope drive and allowing the moon to drift
across the detector (Fig. 3). The motion of the moon in declination resulted in the vertical displacement of subsequent
drift scans from which thermal contour maps were constructed.
By comparison, in Fig. 4, a similar scan across the lunar diameter with the faster Ge-Hg photodetector required 5 sec.
In order to employ this rapid-scan technique, it was necessary
to track the moon by observing a small lunar feature while the
infrared detector moved in the focal plane across the lunar
image as shown in Fig. 5. The tracking was accomplished with
a moderate-sized guide scope. The telescope drive was set
near lunar rate, and small differential corrections to the
tracking were made during a scanning sequence when necessary.
The rapid-scan technique could be used on the illuminated moon
and during an eclipse because surface features were visible.
On the dark side, features were not usually visible, and so a
modified technique is used. A visible feature was located
near the terminator, and the telescope either drifts or is

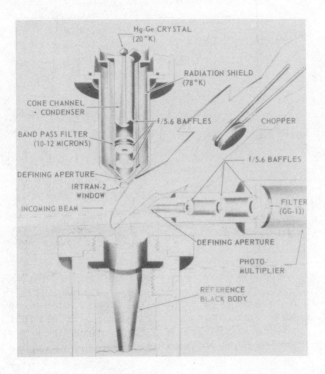

Fig. 2 Cutaway of the instrument used by Shorthill and Saari[14] showing the photomultiplier and mercury-doped germanium detector assemblies, the blackbody reference source, and reflecting chopper.

driven across the terminator onto the dark side. If the dark limb of the moon can be observed, it aids in the determination of the track across the disk.

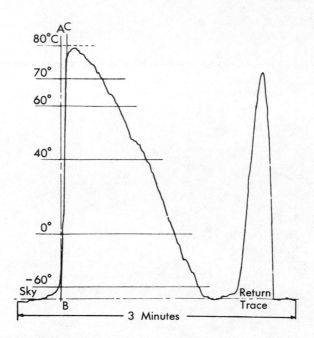

Fig. 3 A sample of a scan line from a lunar map
made by Sinton.[10] His method of obtaining
the assumed limb and the intercept with the
temperature levels are shown by the line AB.

C. Data Recording

Pettit and Nicholson read deflections from a galvanometer
requiring about 5 min for each reading. Various techniques
are now employed for recording, depending on the type of data
analyses to be done. Simultaneous chart records, FM magnetic
tape recordings, and punched tapes have been used. Salisbury[17] recorded his thermal measurements in the form of reconstructed images. For measurement over the entire disk, the
data were recorded on magnetic tape or punched tape, and computer data reduction techniques were employed. Typically, 200
scan lines over the full moon at 10 sec of arc resolution with
appropriate digitization produced 200,000 data points.

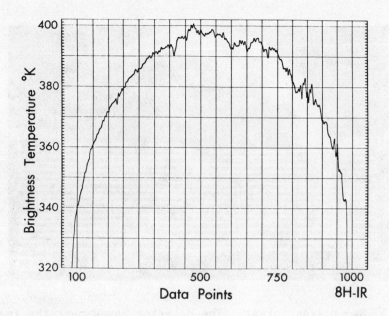

Fig. 4 Rapid scan line across the full moon. The
aperture corresponds to 4.6 data points.
Two hundred scan lines were required to cover
the disk.

D. Location Identification

When the telescope and detector are collimated, identifica-
tion of the region being measured can be accomplished by observing visually or by taking photographs. If the telescope
is moving or if the detector is moving, it becomes more difficult to relate recorded signal variations to given surface
features. In principle, if lunar rates, telescope rates,
scanning rates, and directions, along with at least one surface
feature on the record, are known, then positions can be determined. Most observers have used the bright features and/or
terminator signal to orient their maps.

Shorthill and Saari have used a different method.[15] When the
moon was scanned, neither position nor rotation of the raster
pattern relative to the moon was accurately known. Infrared
and visible radiation were recorded simultaneously, later
digitized, and photometric and isothermal contours constructed
for the visible portion of the lunar disk. Small lunar features could be identified on the photometric contours by their

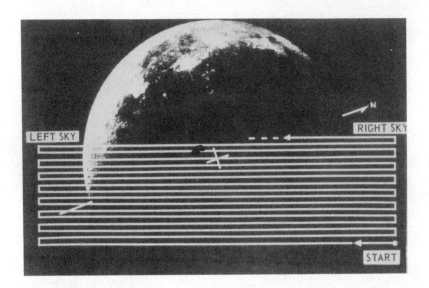

Fig. 5 Rapid scan format. The detector moves over the image
        in the focal plane (every fifth scan line shown in the
        figure).

contrasting albedo, by their shadows, by their corresponding
infrared signature, and by using photographs taken after each
scan program. A transformation was determined by a least-
squares fit between the observed scanner coordinates of these
small features and an orthographic ξ,η librated grid to correspond to the topocentric position of the moon at the time of
the scan. New scanner coordinates were determined for each
feature by using the transformation. The rms value of the
residuals between observed and calculated position was determined to be a fraction of the resolution element.

E. Measurements

1. *Lunar daytime*. The first survey of lunar daytime temperatures was done with a resolution of 25 sec of arc (1 sec
of arc corresponds to approximately 1/1800 of the lunar diameter) by Sinton.[10] He used a narrow-band filter at 8.8 μ
and constructed isothermal maps at nine different phases for
the illuminated portion of the surface.

A more extensive study of lunar daytime temperature
using a 10- to 12-μ filter was done by Saari and Shorthill[15]
for 23 phase angles from -125° to +135° (including simultaneous

measurements in the visible). With a resolution of 8 and 10
sec of arc, the isothermal contours could easily be related to
surface features. The standard orthographic grid lines are
drawn for every 0.1 of the lunar radius. In another set of
measurements, Salisbury[17] scanned the lunar disk at several
different phases and produced images.

2. *Lunar eclipse.* The measurements of Pettit,[18] Sinton,[4]
and the survey of a limited number of ray craters during the
Sept. 5-6, 1960 eclipse by Saari and Shorthill[19] provided a
qualitative indication of the surface cooling. During the
eclipse of Dec. 18-19, 1964, however, scans of the entire disk
at 10 sec of arc resolution were made at four different times
during the postpenumbral phase and three times during totality.
Both thermal contours and reconstructed thermal images were
produced.

3. *Lunar nighttime.* The post-sunset cooling of the lunar
surface is not well determined. Measurements have been made
by Shorthill and Saari[14] and Murray and Wildey[20] using a mercury-doped germanium photodetector. The Surveyor spacecraft
also made post-sunset measurements.[21] Figure 6 shows examples
of lunar nighttime cooling. Murray and Wildey[22] made scans

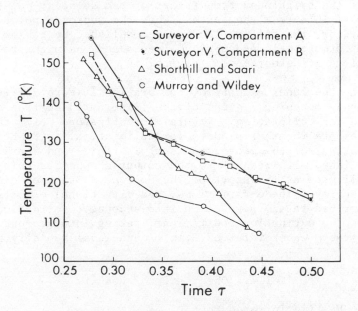

Fig. 6 Lunar nighttime cooling for a region
near the lunar equator. τ is fraction
of lunation; sunset is at 0.25 τ.

over the darkside and produced a thermal contour map of the
nighttime portion for the third quarter moon. Post-midnight
measurements are especially difficult, requiring very sensitive
instrumentation. Low[23] and Mendell and Low[24] used a germanium
bolometer incorporating an interference filter with a 17.5-μ
wavelength cut-on to make repeated scans across the unilluminated moon and reported predawn surface temperatures of approximately 90°K.

F. Calibration

The absolute calibration of an infrared system through the
atmosphere includes a number of difficulties,[25] the most
troublesome being the losses in the optical system and in the
atmosphere. It is possible to determine the optical losses
with an extended blackbody calibration source external to the
telescope[26]; however, there are practical difficulties with
large telescopes. Determination of atmospheric losses depends
upon a knowledge of atmospheric composition (i.e., water
vapor), pressure, temperature with altitude, the zenith angle,
and the extinction law as a function of wavelength.

Shorthill and Saari[15] calibrated their data using a calculation of the brightness temperature of the subsolar point when
it was visible from the earth. When the subsolar point was on
the farside, the calibration was accomplished by taking into
account atmospheric and optical losses and the night-to-night
variations of detector sensitivity.

During the lunar nighttime, calibration is more difficult.
Previously measured regions of the illuminated surface could
be used for calibrating. Several investigators refer their
measurements to other planets for calibration.[22,23] Eclipse
calibration for the measurements of Shorthill and Saari[27] was
accomplished with respect to the computed subsolar point on
the full moon at the start of the eclipse. Sun-moon distance,
local surface albedo variations, and directional effects (to
be described later) were taken into account. Corrections were
made to the extinction coefficient, taking into account the
precipitable water inferred from ground-level humidity measurements.

III. Observational Results

A. The Daytime Infrared Moon

Under conditions of solar illumination, the temperature of
the lunar surface is determined by the solar constant, the
angle of the sun to the local surface, and the albedo of the

surface. With sufficient resolution, the variations caused by
local slopes and local albedo can be observed. The northwestern interior slopes of the crater Theophilus ($\xi = 0.435$,
$\eta = -0.198$) are cooler because at the local surface the elevation angle of the sun is smaller, whereas the southeastern
slopes are warmer because the local surface is more normal to
the sun. Mare Nectaris ($\xi = 0.55$, $\eta = -0.25$) southeast of
Theophilus is darker (lower albedo) than its surroundings and
is therefore elevated in temperature by about 3°K. These effects can be seen in a thermal image (see Fig. 7) reconstructed from the same data used to produce the isothermal contours.
In the image, the region N is Mare Nectaris and T is the
crater Theophilus. Many other features can be identified in
the image; bright tones represent warm regions, whereas the
dark tones indicate cooler regions.

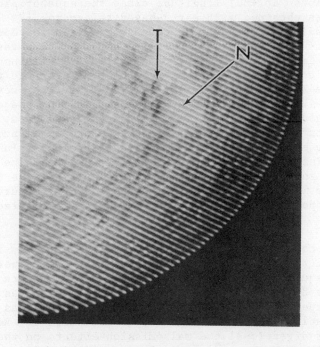

Fig. 7 Infrared image of a full moon scan.
The data were obtained at phase of
$-2°$ $16'.1$ at 23^H 34^M3 UT on
Dec. 18, 1964. The moon was scanned with 200 lines. The every-other-
line effect is caused by the electronic enhancing of the image.

Simultaneous isothermal and isophotic contour maps[15] have
been constructed for 23 lunar phases from -125° to +135°.
Position information is given by the orthographic projection
of the librated selenographic latitude-longitude or standard
grid systems. The position error in the various grid fits was
about 2 sec of arc. A portion of one of the isothermal contour maps is shown in Fig. 8 for the southwest quadrant, along
with calibration data. The simultaneous photometric maps are
useful for study of the relation between albedo variations and
brightness temperatures. An example of the isophotic contours
is shown in Fig. 9, where the contours were used for the construction of an albedo map in relief. The region is Mare
Tranquillitatis, and the feature being pointed out is the
bright crater Plinius.

The radiometric lunation curves for the center of the moon
and for points   at +30° longitude along the equator plotted in
Fig. 10 were obtained with earth-based telescopes.[14] The successful landing of Surveyor spacecraft provided the opportunity
of making *in situ* temperature measurements. A temperature
sensor was imbedded in the outboard face of a well-insulated
electronic compartment. It was possible to infer the temperature of the local surface during the lunar day, for several
days after sunset, and during an eclipse. During the daytime,
there was general agreement with the earth-based data; both
sets of data show a higher temperature before local noon,
which is to be expected from directional effects of lunar
thermal emission. The first results obtained from the Surveyor
series are shown in Fig. 11. Detailed results of the Surveyor
temperature measurements have been reported by the Surveyor
Thermal Properties Working Group[28-30] and are summarized in
Chap. 2a.

The moon does not radiate infrared energy uniformly in all
directions. Pettit and Nicholson found that the temperatures
along the equator at full moon vary as $\cos^{1/6}\theta$ (where θ is
the angular distance to the subsolar point) and not as
$\cos^{1/4}\theta$ expected from a Lambertian surface. This was qualitatively explained by the roughness of the surface. Recent
analysis of directional thermal emission effects on the full
moon reveals that $\cos^{1/6}\theta$ is only an approximation to the
experimental values, particularly near the limb. The variation of brightness temperature over the disk of the full moon
has been studied by considering the data statistically. a
least-squares fit to the statistical data is represented by
the equation $T(K) = 324.2 + 72.6 \cos\theta$. In Fig. 12 the statistical distribution of brightness temperature for the full
moon is compared with the $\cos^{1/6}\theta$ variation. The variation

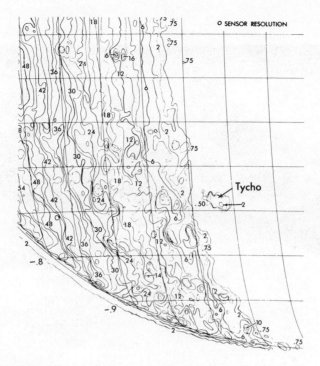

QUADRANT III   PHASE ANGLE 112°

Contour Number	Temperature °K	Contour Number	Temperature °K	Contour Number	Temperature °K
.50	143.9	30	260.7	62	304.2
.75	150.7	32	264.1	64	306.4
2	169.7	34	267.3	66	308.6
4	186.4	36	270.5	68	310.7
6	197.7	38	273.5	70	312.8
8	206.7	40	276.4	72	314.9
10	214.2	42	279.2	74	317.0
12	220.7	44	282.0	76	319.0
14	226.6	46	284.7	78	321.0
16	231.9	48	287.3	80	322.9
18	236.8	50	289.9	82	324.9
20	241.4	52	292.4	84	326.8
22	245.7	54	294.8	86	328.7
24	249.7	56	297.2	88	330.5
26	253.6	58	299.6	90	332.4
28	257.2	60	301.9	92	334.2

Fig. 8   Isothermal map of the southwest quadrant of the illuminated portion of the moon at phase of 111° 40'.1 for
13^H 00^M UT on Oct. 11, 1963. Note that the crater
Tycho is 170°K one day after sunset, and its environs
are less than 144°K.

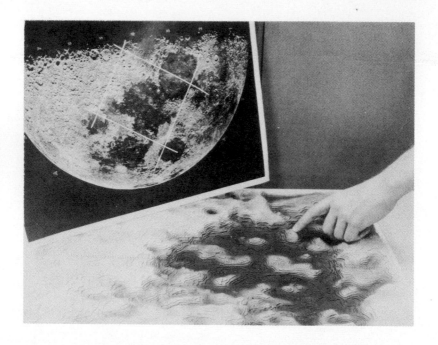

Fig. 9   Relief representation of the full moon iso-
         photic contours over the region of Mare Tran-
         quillitatis. The crater indicated is Plinius.
         Other features may be identified using the
         lunar photograph.

of brightness temperature at full moon in the N–S direction
from the disk center is shown in Fig. 13. Also shown in
Fig. 13 is the variation in brightness temperature along the
equator for the first and third quarter moon. The theoret-
ically expected $\cos^{1/4}\theta$ variation follows the observed data
for quite a distance from the subsolar point, but as the terminator is approached the temperature rises above this theoretical value. An extensive study was performed using the
isothermal contour maps through a lunation to obtain a better
understanding of the directional characteristics of thermal
emission. It was found to depend in a rather complicated way
on the elevation angle of the sun and the observer's azimuth
and elevation angles.[31] The data were studied in terms of a
"thermal coordinate" system, where the subsolar point is taken
as the north pole (90° thermal latitude) and the terminator,
the equator (0° thermal latitude). Thermal longitudes, great
circles through the pole, are measured counterclockwise around

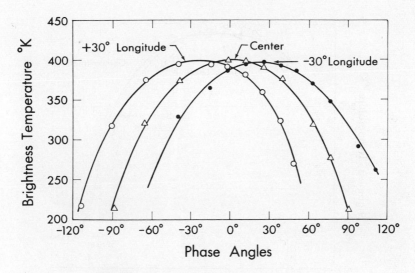

Fig. 10 Radiometric lunation curves for three points
along the lunar equator.

the subsolar point, zero longitude (the thermal meridian)
passing through the disk center. Figure 14 shows the grid
system for a scan made at a phase angle of 25° 37' after full
moon. Brightness temperatures were measured every 10° in
thermal latitude and longitude; along a particular latitude
where the sun angle is constant, the temperatures, in the absence of directional effects, should be constant. The data
from this phase are shown also in Fig. 14, where little
variation in temperature is seen for high latitudes. In a
special study, the brightness temperatures along the thermal
meridian for sun elevation angle every 10° were corrected for
the sun-moon distance and for albedo variations. These data
were then converted to elevation angle of the observer with
respect to the surface, and a least-square polynomial fit was
made for the values. For an observer on the surface, the 90°
observer elevation angle corresponds to looking down vertically
at the surface, 0° to looking at the horizon with the sun to
his back, and 180° looking at the horizon into the sun. This
is shown in Fig. 15, for sun elevation angles from 10° to 30°,
the highest temperatures, are obtained on the horizon, whereas
for the larger sun elevation angles the highest temperatures
are obtained at observer angles somewhat lower than the sun.
Since these data were taken along the thermal meridian, the
line of sight is in the plane defined by the surface normal
and the sun. The directional factor with the sun at 60° elevation can also be plotted in terms of an azimuth-elevation
coordinate system; the azimuth is zero in the direction of the
sun. The directional factor is defined as the ratio of the

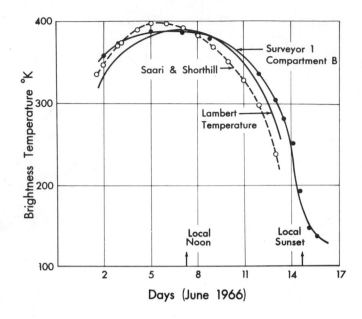

Fig. 11   Lunation temperature curves for the
          Surveyor I site. The Lambert tem-
          perature is the temperature that a
          perfectly insulating Lambert surface
          with unit emissivity would come to
          if it absorbed the same amount of
          radiation as the surface under con-
          sideration.

observed brightness temperature to the Lambert temperature of
the surface. This factor was determined over the lunar disk
at points where the sun was at 60°, and the appropriate azimuth
and elevation were calculated. A spherical harmonic fit to
these data points is shown in Fig. 16. The highest brightness
temperature (4.3% above the Lambert temperature) is obtained
for 0° azimuth (looking toward the sun) at an elevation of 47°,
somewhat below the sun. The lowest temperature is found at
180° azimuth and 0° elevation. In general, high values of
brightness temperature are obtained when the angle defined by
the sun and emission being measured (lunar phase angle) is
small and low when the phase is large. Another example of the
directional factor is shown in Fig. 17, which is plotted in
terms of the sun elevation angle along the thermal meridian
and the observer's elevation angles. The plot of the subsolar
point variation with phase is, therefore, the upper boundary
of the figure (sun at 90°), and the lower boundary is the
(sun at 0°). The directional data at various sun angles were

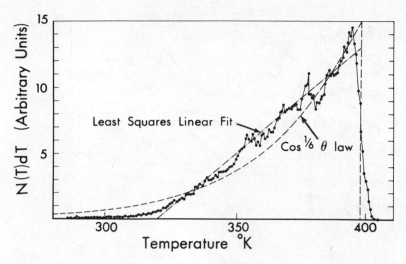

Fig. 12   Statistical distribution of brightness tem-
          peratures for the full-moon disk. Fluctua-
          tions in the distribution are caused by
          variations in the average albedo.

used in order to interpret the Surveyor temperature measurements[29] when the lunar surface was illuminated by the sun.

B. The Nighttime Infrared Moon

   The thermal response of the lunar nighttime surface is not
well determined at this time. An example of data used to infer the midnight temperature is shown in Fig. 18, along with
direct measurements of the antisolar point. The data are
plotted in terms of days after local sunset. There are differences in the extrapolated equatorial scan data, which may
be caused by drifts in the sky background. Also shown is
$(122 \pm 3)°K$, the Shorthill and Saari value of 99°K, and the
Low[23] value of 90°K. The value of 106°K was determined by ex-
trapolating the Murray, Wildey[20] data. The Pettit and Nicholson value was corrected by Saari[32] to 108°K. Ingrao[33] has
suggested that the curve of Murray and Wildey should be shifted
by 5 hr toward the terminator. The extrapolation to the midnight value then becomes 104°K and was shown shifted that way
in the previous Fig. 6, along with the nighttime results from
Surveyor V, which gives a value of 116°K.

   An isothermal map of the unilluminated part of the third
quarter was constructed by Wildey, Murray, and Westphal,[22]
where they reported over 100 nighttime anomalies, most of
which correspond with hot spots found on the eclipsed moon.

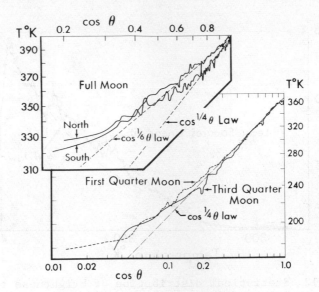

Fig. 13 Variation of brightness temperature in the
         N-S direction from the disk center are taken
         from the full moon scan at phase angle of
         -2° 16'.1. The lower graph shows the
         temperature in the E-W direction along the
         equator at first and third quarter moon.
         Fluctuations in the data are caused by
         variations in the average albedo and shadow-
         ing.

Two examples of scan line data across a nighttime anomaly are
shown in Fig. 19, plotted in terms of hours after sunset. The
general cooling of the environs of Copernicus can be seen, as
well as the anomalous cooling of the crater itself. The area
of thermal enhancement extends beyond the crater, as it does
during an eclipse. However, it is surprising that there is no
evidence of thermal structure in the crater or on the rims, as
there is during an eclipse. The crater Tycho, however, does
show evidence of nighttime thermal structure in the infrared
images of Hunt, Salisbury, and Vincent.[34] With the rapid scan
system of Saari and Shorthill,[15] isothermal maps of Tycho were
obtained for three different times after sunset. The previous
Fig. 8 shows Tycho about 12 hr after sunset, and Fig. 20 shows
Tycho about 24 hr after sunset. Selenographic grid lines are
superimposed, and the crater center is indicated with a cross.
The lunation temperature curve for Tycho was determined from
the isothermal contour maps.[15] The brightness temperature departs somewhat from the insolation because of directional

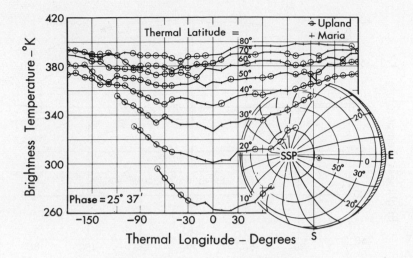

Fig. 14   Brightness temperature vs thermal longitude for
lunar phase 25° 37'. Variations are caused by
albedo differences and shadowing. The thermal
coordinate system for this phase is shown in
the lower right corner.

effects explained previously and is plotted in Fig. 21. The
three values following sunset are reliable. The two presunrise
data points cannot be considered as reliable, since they were
obtained by averaging all of the data points within the crater.
Post-midnight temperature measurements require extremely
sensitive instrumentation. Low,[23] however, using the 17.5- to
22-µ window and cancelling the sky background with a differential method, has measured temperatures as low as 70°K on
the lunar surface. Figure 22 shows a typical scan of the derivative of the intensity profile across the nighttime surface.
The bright limb and terminator show off-scale deflections.
Thermal structure is indicated by the variations in the signal.
Note that the sky noise has been almost completely eliminated.
The point labeled "hot spot" reported by Low near the southeast limb has been tentatively identified from the eclipse
data as Schomberger F.

C.   The Eclipse Infrared Moon

Perhaps the best summary of the observational results obtained for the eclipsed moon is represented by an image reconstructed from the scan line data of Saari and Shorthill.[35]
Although images represent the data in only a qualitative
manner, the major results are indicated. Figure 23 shows the
thermal emission from the moon during a total eclipse at a

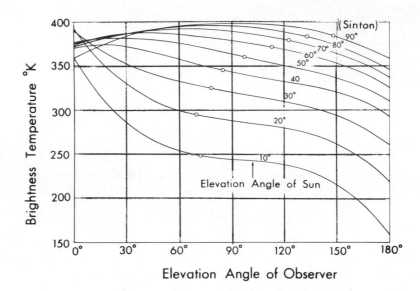

Fig. 15 Brightness temperatures vs observer's elevation
          angle to the surface with the sun at different
          elevation angles. The data for 90° sun angle
          are taken from the observations of the sub-
          solar point made by Sinton.[4] The Lambert
          temperature is indicated with an open circle.

resolution of 10 sec of arc (1/200 the lunar diameter). The
north line is inclined at 30° to left of center in the upper
left portion of the figure. The bright spots number about
1000 and have been referred to as "hot spots" in the literature. The large spot 1/5 the way from the bottom is Tycho;
the other large spot to left of center is Copernicus. In
addition to these two ray craters already known to be thermally
anomalous, all the major ray craters, such as Aristoteles,
Dionysius, Langrenus, Menelaus, and Stevinus, are also anomalous. Besides these localized hot spots, extended regions of
thermal enhancement can be seen on the image which are related
to certain maria and portions of maria. One example is Mare
Humorum to the left of Tycho.

    The normalized eclipse cooling curve for the center of the
lunar disk is shown in Fig. 24 plotted in terms of the eclipse
time normalized to the penumbral duration. The curve is not
significantly different from that obtained by Pettit in 1939.
An eclipse cooling curve for any point on the earthside hemisphere may be obtained from the data of Shorthill and Saari.
Isothermal contours have been constructed for the seven eclipse
scans and for the full moon before the eclipse. The isothermal

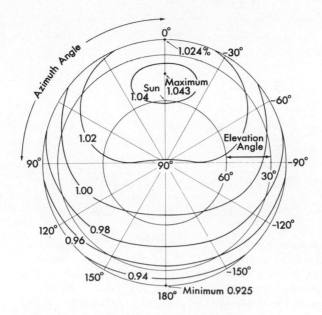

Fig. 16   Contours of directional factor for the sun at
          60° elevation. The directional factor is de-
          fined by the ratio of observed temperature to
          that of the Lambert temperature.

contours obtained at full moon just prior to onset of the
eclipse for the crater Tycho are illustrated in Fig. 25. A
lunar Orbiter photograph of Tycho in Fig. 26 has the selenographic latitude and longitude superimposed. (The photograph
is not at full moon.) The full-moon temperature of Tycho is
slightly below its environs because of its relatively high albedo. As Tycho cools during the eclipse, the distribution of
brightness temperature changes. In fact, a region around the
southeast interior which was the coolest part of the crater at
full moon (359.6°K), which is 9.1°K below its environs) is the
warmest part during totality (227.1°K, which is now 51°K above
its environs). The eclipse isothermal contours for the Tycho
region are shown in Fig. 27, with the anomalous region extending beyond the crater by about half a crater diameter. There
are three relative maxima seen in the thermal structure, one
near the central peak, one along the west interior, and one on
the southeast floor or interior wall. Copernicus has similar
characteristics. Detection of thermal structure in smaller
craters was limited by the resolution used in the measurements.

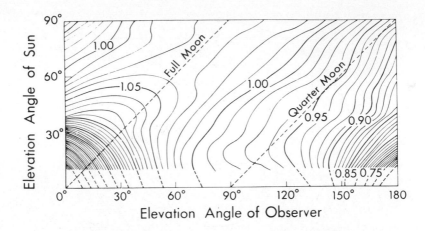

Fig. 17 Contours of directional factor for points on the thermal meridian (the great circle through the disk center and the subsolar point). The directional factor is defined as the ratio of observed temperature to that of the Lambert temperature.

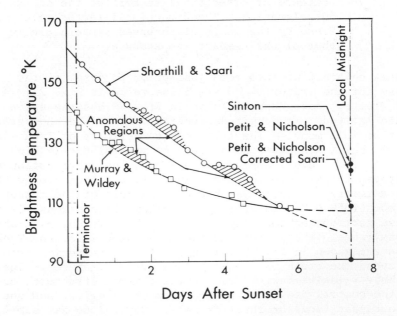

Fig. 18 The equatorial brightness temperature in terms of days after local sunset. The anomalous regions are due to maria or hot spots.

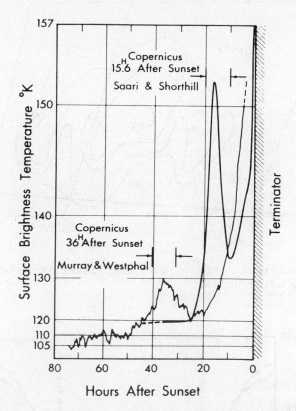

Fig. 19   Lunar nighttime temperature trace
          through the crater Copernicus.
          The data of Shorthill and Saari
          were replotted and appear abnormally
          smooth.

The eclipse cooling curves for Tycho and its environs are
shown in Fig. 28. The data point at normalized time 3.5 should
be considered tentative because of certain calibration problems.
The points plotted in this figure were obtained from the
average of about 50 data points reading across the crater.
During the penumbral phase of the eclipse, the cooling curves
depart from the insolation curve because of the thermal inertia
of the surface.

A special study was done on 83 of the hot spots with the
largest signal difference over their environs.[36] Many of these
were associated with craters smaller than the projected sensor
diameter. An areal correction was made with the assumption
that the hot spot was confined to the crater and the emission

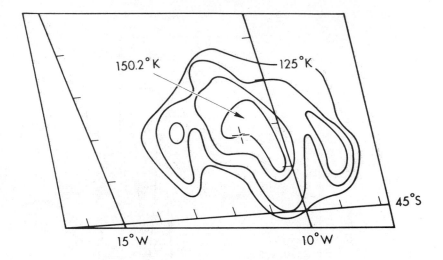

Fig. 20   Isothermal contours over the crater Tycho about one
day after sunset.

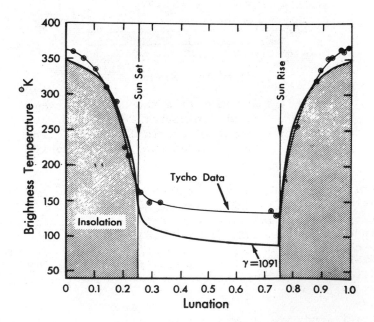

Fig. 21   Lunation temperature curve for the crater Tycho.
The curve departs from the insolation because of
directional effects.

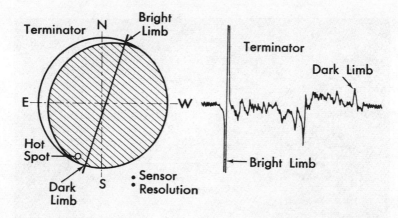

Fig. 22   An example of a scan across the nighttime
          surface showing the derivative of the in-
          tensity profile at 20 μ.[23] The hot spot
          reported by Low has been tentatively
          identified as Schomberger F.

was uniform across the crater. The correction took into account the crater diameter, its projection onto the sensor diameter, and the relative signal difference above the environs
on successive scan lines. The 30 most prominent areally corrected craters were ranked and are shown in Table 1. Six are
ray craters, 20 are craters with bright interiors, and 4 have

Table 1   Ranking of 30 prominent anomalies on the eclipsed moon

No.	Crater	Diam, km	No.	Crater	Diam, km
1	Mösting C	3.8	16	Egede A	12.5
2	Piton B	4.8	17	Laplace A	9.7
3	Messier A	13.6	18	Nicollet	15.2
4	Bunch B	6.7	19	Mösting A	13.0
5	Jansen E	7.0	20	Mason C	12.3
6	Torricelli	6.9	21	Cauchy	12.3
7	Draper C	7.7	22	Gambart C	12.2
8	Maraldi B	7.4	23	Carlini D	9.3
9	Moltke	6.4	24	Eudoxus A	14.1
10	Plato M	8.3	25	Pico B	11.4
11	Guericke C	10.9	26	Cephus A	12.5
12	Flamsteed B	9.4	27	Hesiodus B	10.2
13	Taruntius H	8.3	28	Janssen K	15.5
14	Jansen F	9.4	29	Bode A	12.3
15	Marius A	16.0	30	Carlini	11.4

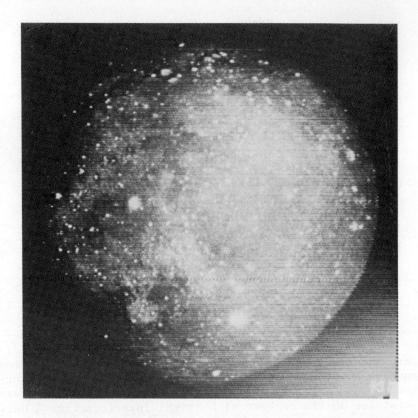

Fig. 23   Thermal image of the eclipsed moon during totality.
          The midpoint of the scan is 2^H $18^M_.8$ UT, Dec. 19,
          1964. Individual scan lines can be seen. The
          bright region in the lower right corner is caused
          by radiation from the side of the telescope.

bright rims. There are 5 in the uplands and 23 in the maria.
A plot of areally corrected signal difference vs crater diameter is shown in Fig. 29. It should be noted that most
(68%) of the hot spots are smaller than the sensor diameter.
The plot represents the boundary of a general distribution for
a large number of hot spots. It is expected that as the resolution is improved a larger number of hot spots will be observed. There is an upper limit on the signal difference over
the environs if the thermal parameter γ is not less than 20
(bare rock). Mösting C was observed to have a $\Delta T = 28°K$
above its environs, whereas the areally corrected temperature
difference, $\Delta T_C = 157°K$, suggests the properties of bare
rock. The Lunar Orbiter spacecraft photographed certain of
the hot spots with high resolution. It is evident from these

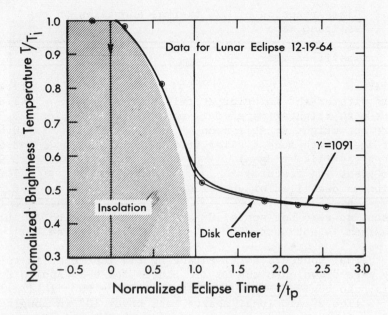

Fig. 24   Eclipse cooling curve normalized to initial
temperature and penumbral duration. The
constant properties model plotted here
assumes constant k (thermal conductivity),
constant c (specific heat) with temperature,
and constant ρ (density) with depth.

high-quality photographs that these craters are among the
younger features on the lunar surface. In general, they appear to be rougher and to have a higher number of boulders inside and around the rims than older craters. The Lunar Orbiter
photographs of Mösting C (Fig. 30) show that the crater has
sharp features with an interior looking like bare rock and the
ejecta blanket covered with boulders.

A classification of the 330 hot spots revealed that most of
them are associated with regions or craters with high albedo.
Table 2 lists the preliminary finding; of those listed, two-
thirds fall in the lunar maria and one-third in the uplands.
The density distribution of all the hot spots over the hemisphere is shown in Fig. 31.[37] The greatest concentration is
found in Mare Tranquillitatis and in a region between Kepler
and Aristarchus. Several different randomness tests showed
that the Mare Tranquillitatis concentration occurring by
chance was 1 in 5000. The contour interval is 0.7 hot spots/
10^5 km^2. The mean in the distribution is 5.34. The minimum,

Table 2   Classification of 330 thermal anomalies on the
          eclipsed moon

Classification	%
Ray craters	19.4
Craters with bright interior at full moon	41.8
Craters with bright rims at full moon	23.3
Craters not bright at full moon	0.6
Bright areas with much smaller crater	3.6
Bright areas (like ridges)	3.9
Bright areas (no feature)	1.2
Position unidentified or questionable	6.3

representing zero hot spots/10^5 km^2, is labeled with an L;
the maximum 14 hot spots/10^5 km^2 is labeled with an H.

The thermal contours obtained during the last scan in totality were drawn onto the 20 AIC charts for the "Apollo Band,"
that is, the region $\pm 8°$ in latitude and $\pm 50°$ in longitude.[27] (The 20 AIC lunar charts each cover 10° in longitude
and 8° in latitude and were constructed by U.S. Air Force Aero-
nautical Chart and Information Center, St. Louis, Mo.) The
original contours were drawn by computer, along with the appropriate librated selenographic grid in longitude and

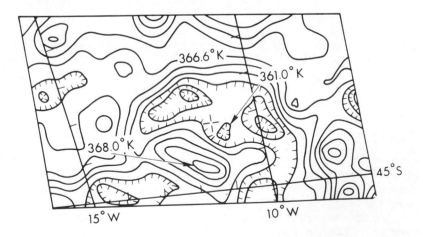

Fig. 25   Full-moon isothermal contours just prior to an
          eclipse. The northwest interior walls are war-
          mer because the sun is more normal to the sur-
          face; similarly, the northeast interior walls
          are cooler because the sun is more grazing.

Fig. 26   Lunar Orbiter photograph of the crater Tycho. The
          thermal contour shown in Figs. 20, 25, and 27 can be
          compared to the features on this photograph.

latitude. The formal rms error in the librated grid was
2.3 sec of arc or about one-quarter of the sensor resolution.
In the grid-fitting process already described, 45 known hot-
spot features distributed over the disk were used. The contours were manually transferred onto each AIC chart. An example of one such chart is Wichmann AIC 75B 40° to 30° W
longitude, 0° to 8° S latitude (Fig. 32). A description follows: there are relatively few hot spots in this region of
Oceanus Procellarum. The most prominent is Lansberg D
(30°.6 W; 3°.0 S) with a $\Delta T = 27.5°K$. The extent of the contours in the scan-line direction suggests that the hot spot is
confined to or within the crater. Lesser hot spots are observed on Lansberg E (30°.3 W; 1°.8 S) with a $\Delta T = 10.5°K$,
Wichmann (38°.6 W; 7°.6 S) with a $\Delta T = 8.1°K$, and Wichmann B
(39°.6 W; 7°.1 S) with a $\Delta T = 8.2°K$. In this chart, the thermal maxima for hot spots associated with craters all seem

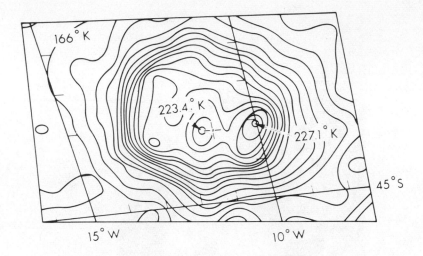

Fig. 27 Isothermal contours over the ray crater Tycho during
         the totality of the lunar eclipse of Dec. 19, 1964 at
         3^H $2^M.8$ UT. The anomalous region extends beyond the
         crater by at least half a crater diameter.

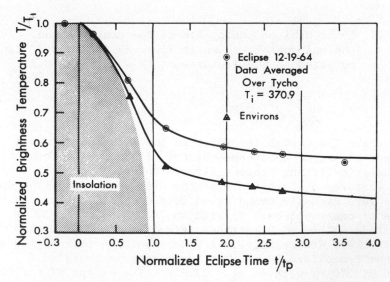

Fig. 28 Normalized eclipse cooling curve for the ray crater
         Tycho. The data are averaged over the crater for
         each plotted point.

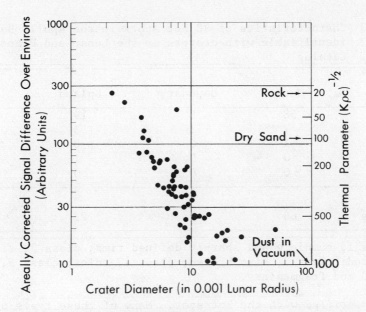

Fig. 29   Distribution of hot spots vs crater di-
          ameters. The resolution of the sensor
          corresponds to 0.01 of the lunar radius.
          The signal difference, when compared to
          the theoretical cooling of a constant
          properties model of the lunar surface,
          corresponds to values of the thermal
          parameter, $\gamma = (k\rho c)^{-1/2}$.

to be displaced somewhat to the west. A mild hot spot with a
$\Delta T = 13.0°K$ at 35°5 W; 4°8 S is not associated with a crater.
It also appears as a slight enhancement in the MIT 3.8-cm
radar image; the Lunar Orbiter IV frame 137 photograph shows a
number of kilometer-sized craters in this region.

There were 165 hot spots found in the Apollo Band, the majority of which had a $\Delta T > 5°K$ over their environs. The
average $\Delta T = 12.3°K$. The actual corrected average T_c will
be different, since many of the features are smaller than the
sensor diameter and must be areally corrected. The hot spots
identifiable with craters listed in the University of Arizona,
Lunar and Planetary Laboratory Catalog of lunar craters are
given in Table 3 according to their class and general location.
The table shows that 60% of the hot spots in the Apollo Band
associated with class 1 are craters and occur in the maria.
In the Apollo Band as well as in other regions on the lunar
surface, so-called "no-feature" hot spots are found. In these
regions, there is no obvious feature or crater of significant

Table 3   Characteristics of 98 hot spots in the Apollo Band
          identifiable with craters in the Lunar and Planetary
          Catalog

Crater[a] class	Mare	Boundary	Upland	Class total
1	58	8	19	85
2	7	1	3	11
3	1	0	0	1
4	0	0	1	1
5	0	0	0	0
Total craters	66	9	23	98

[a]Class 1, complete and sharply defined rims; class 2, rims
blurred; class 3, rims broken; class 4, ruins; class 5, bat-
tered and fragmentary.

size identified with the hot spot. Many of these types of hot
spots show radar enhancement and from the Lunar Orbiter photography are often associated with a very small bright crater or
cluster of small bright craters and in some cases with regions
or "patches" of high albedo.

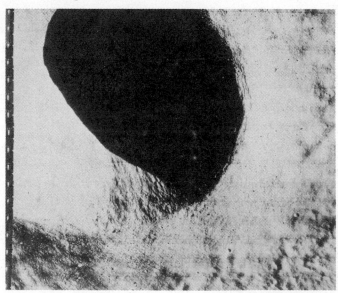

Fig. 30   Lunar Orbiter photograph of Mösting C
          (8°1 W, 1°8 N). The crater diameter
          is 3.84 km.

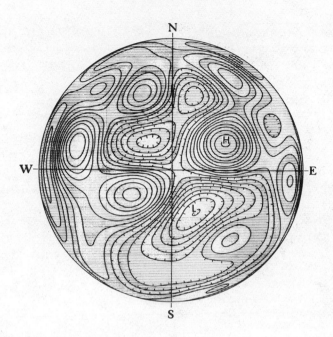

Fig. 31   A spherical harmonic fit to the density of
          1011 hot spots on the lunar hemisphere with
          a contour interval of 0.7 hot spots/10^5 km^2.
          The mean in the distribution 5.34. A high
          of 14 (indicated by H) is located in Mare
          Tranquillitatis. The minimum representing
          zero hot spots/10^5 km^2 is labeled with an
          L.

Because of their unusual appearance, they may indicate a different formation process. Figure 33 shows such a feature in
the ray system of Kepler at 42°90 W; 7°08 N. This feature was
detected on two successive infrared scan lines. The two peak
temperatures, as well as the environs, were plotted. It is apparent that the peak temperature depends on just how the sensor
passes through the hot spot (see Fig. 34). An area correction
was made on the assumption that the source was 18 km in diameter. Since the magnitude of the two signals on successive
scan lines depends on the geometry of intersection of the sensor and hot spot, the temperature of the "feature" can be determined. The areal correction produced a smooth cooling curve
that is parallel to the environs curve.

Radar images in the equatorial band have been constructed by
the MIT Lincoln Laboratory.[38] These measurements made at
3.8 cm show some remarkable correlation with the infrared data.
Their resolution was substantially better than any of the

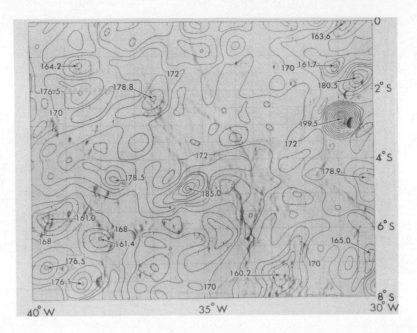

Fig. 32   Isothermal contours on the AIC base. Charts
          were drawn originally to a scale of 1:500,000
          and have been completed with 20 regions ± 8°
          latitude and ± 50° longitude. The region at
          35°.5 W longitude and 5°.3 S latitude is a hot
          spot for which no related visible feature can
          be identified.

infrared data. No radar features were listed closer than 23°
from the center of the lunar disk, for reasons described in
the foregoing reference. When this difficulty is removed, additional correlation should be expected within 23° of the disk
center. Table 4 lists the 3.8-cm radar enhancements and the
corresponding infrared hot spots. In several instances, the
apparent source of the hot spot is a small region of high albedo, sometimes with small craters similar to those in Fig. 33.
In most cases, however, the major radar enhancements listed in
Table 4 can be related to moderate-sized craters. The correlation between the radar enhancements and hot spots is interesting because radar is sensitive only to structure of the surface
and not to thermal emission. For example, a region that is
covered with rocks and/or characterized by exceptionally dense
surface material will exhibit enhanced radar backscattering
and anomalous eclipse cooling. This would suggest that a
method for detecting an internal heat source would be to observe areas that are not radar enhancements but have anomalous
cooling during an eclipse or the lunar nighttime. Considering,

Table 4 Location of relatively large regions of strongly enhanced local roughness from 3.8-cm radar data and related infrared hot spots (ΔT is temperature above environs)

Feature	Crater diam, km	Long., deg	Lat., deg	Infrared ΔT, °K	Radar[a] diam, km	Full moon visual appearance
Rocca Ab	10	-68.4	-12.7	11	30	Bright halo
Near Rocca Fa	4	-66.7	-14.5	8	15	Bright halo
Grimaldi G	12	-64.8	-7.8	14	25	Halo and rays
Siralsis J	12	-59.7	-13.4	16	15	Bright interior
Flamsteed Gc	5	-52.5	-3.2	22	25	Halo and rays
Unnamed Crater	4	-50.3	-0.2	9	15	Small bright spot
Suess	9	-47.9	4.0	32.4	25	Bright interior
Gassendi F	5	-45.2	-14.9	17	30	Bright halo
Cluster small craters	...	-42.3	7.2	18.2	40	Bright halo
Encke X	4	-40.2	1.0	17.1	70	Bright halo
Herogonius Ec	4	-36.4	-12.8	7	40	Bright halo
Reinhold Na	2	-25.7	1.9	20.0	25	No brightening
Censorinus	4	32.7	-0.4	25.5	60	Bright halo and rays
Between Capella C and CA	...	36.2	-6.1	8.8	20	Bright halo
Secchi A	5	41.2	3.3	14.8	15	Bright interior
Messier A	14	46.9	-2.0	46.0	30	Halo and rays
Taruntius K	3	51.5	0.7	26	50	Bright interior
E. rim Lick	2	53.3	12.4	8	20	Bright halo

[a]Region of radar enhancement covers an area much larger than the feature.

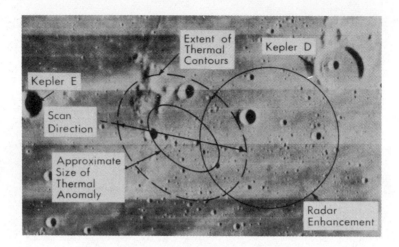

Fig. 33   Lunar Orbiter IV frame 144 photograph. The
          ΔT = 18.2°K above the environs for this hot
          spot located in a ray extending west from
          Kepler. The 3.8-cm radar enhancement was
          reported to be 30 km in diameter. There are
          three craters less than 3 km and numerous
          smaller ones located in the enhanced regions.

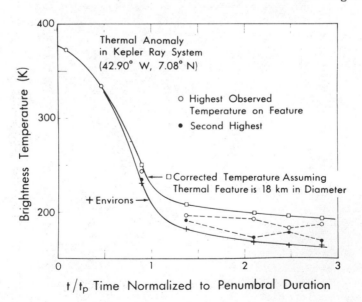

Fig. 34   Eclipse cooling curve for the "no-feature"
          hot spot shown in Fig. 33.

however, the moon's highly insulating layer, Saari[39] has shown
that it is very unlikely that subsurface heat source can be
detected with infrared radiometric measurements. Actually,
the search for internal heat sources could perhaps best be done
by high-resolution microwave radiometry from an orbiting lunar
spacecraft. The observations of lunar transient events described by Middlehurst[40] may be evidence of internal activity.
Many of these events took place in regions that are not associated with thermal anomalies such as Alphonsus, Gassendi,
Plato, and Sabine. A few of these events, however, did occur
in regions that are thermal anomalies, such as Tycho, Aristarchus, and Messier A. There does not seem to be any strong
correlation between transient lunar events and the eclipse hot
spots. It is possible that most of the hot spots are produced
by concentrations of rocks on the surface, in which case the
eclipse measurements are in reality a "rock detector." Lunar
Orbiter, Surveyor, and Apollo photographs show that rocks are
a common feature of the lunar landscape, occurring in and
around craters, on certain ridges, and in some of the rilles.

It is somewhat difficult to compare different areas over the
moon on an equal basis using the eclipse isothermal contour
charts for two reasons. First, the isotherms show a dropoff
toward the limb because of a lower initial temperature T_i,
and second, the temperatures in the west are lower in the east
because westerly points have cooled longer. Figure 35 shows
some preliminary results of removing these two effects to produce a "flattened moon." The trace passes through Mare Humorum
between data points 340 and 400, through Heinsius A to data
point 560, and through Tycho between data points 590 and 640.
Similar data were obtained for the entire set of 200 scan lines
and a contour map of $\Delta T/T_i$ constructed for further study.

IV. Theoretical Model

Previous theoretical models describing the temperature variation of the typical lunar surface were unable to fit simultaneously the lunar nighttime and the eclipse cooling data.
Recently a "particulate model" was developed by Winter and
Saari[12] which fits both sets of data. They describe their
model as a cubic array of cubes; each is in contact with its
immediate neighbor only along common edges, so that the void
fraction is 0.5, and assume particles (cubes) are opaque
throughout the relevant spectral region. They considered each
individual particle approximately isothermal. The effect of
contact conduction is included using the results from Watson's
experiments[41] with evacuated powders. In the model, specific
heat is a function of temperature. Winter and Saari have
plotted the predicted nighttime temperature for the particulate

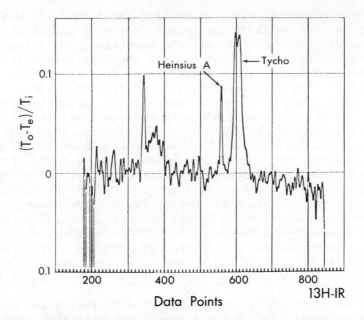

Fig. 35 The trace for the "flattened moon" showing
the ratio $(T_o - T_e)T_i$, where T_o is the
observed temperature, T_e is the empirically determined average lunar surface
temperature, and T_i is the initial full
moon temperature.

model and constant properties model (Fig. 36). (Constant
properties model is characterized by the thermal parameter
$\gamma = (k\rho c)^{-\frac{1}{2}}$ where k is thermal conductivity, ρ density,
and c specific heat.) The curves are compared with observed
lunar nighttime temperatures. For the constant properties
model, the best value for nighttime cooling is a γ of 800,
whereas for an eclipse, a γ of 1300 gives the best fit.

In order to compare the particulate model results with the
experimental data, a new measure of eclipse cooling was de-
vised by Winter and Saari, the "differential energy parameter."
This was necessary so that more meaningful comparisons could be
made, particularly during the penumbral phase of an eclipse.
The differential energy parameter P was defined by the equation

$$P = (T/T_i) - I\,(\tau)$$

In this equation, $\tau = t/t_p$ is time normalized to the penumbral duration starting with the beginning of the penumbra,

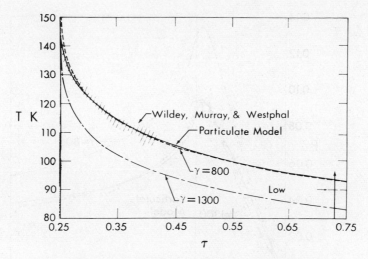

Fig. 36   Theoretical predictions of nighttime
temperatures from the cube model after
Winter and Saari[12] and the constant properties model. Time τ is normalized to
a lunation and is measured from local
noon.

T_i is the full moon temperature at $\tau = 0$, and $I(\tau)$ is the
solar flux normalized to the full moon value at $\tau = 0$. In
Fig. 37, a plot of the predicted eclipse variation is shown for
the particulate model. The black dots are the observed values
from the eclipse measurements of Shorthill and Saari. It can
be seen from the two figures that the particulate model predictions are in excellent agreement with the observed infrared
measurements for both nighttime and eclipse conditions.

The foregoing brief description of the particulate model refers only to the typical lunar surface. We know, however, that
rocks are present on the lunar surface. It is expected that
their thermal behavior during a lunation and an eclipse will
be different from that of the typical lunar surface. Rocks are
observed on the lunar surface from several meters in size to a
few millimeters. Roelof[42] has investigated the thermal behavior of centimeter and meter rocks. It was found by Roelof
that during most of a lunation the centimeter-size rocks are
slightly cooler than the general surface and are approximately
isothermal. During sunrise and sunset, these rocks will have
considerably higher temperatures than the typical lunar surface,
which may account for some of the directional effects (i.e.,
limb brightening at full moon). The thermal response through
a lunation of a one meter basaltic cube on the lunar equator

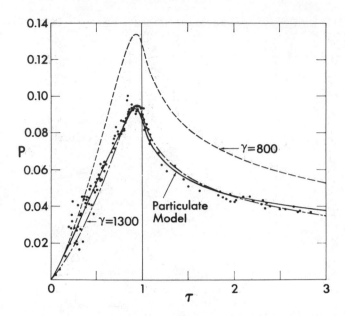

Fig. 37   Theoretical predictions of differential
          energy parameter from the particulate
          model and the constant properties model
          after Winter and Saari.[12] Time τ is
          normalized to the penumbral duration
          and is zero at the start of the penumbra.
          The closed circles are from the eclipse
          data of Shorthill and Saari.

is shown in Fig. 38. It can be seen for the one meter rock
that isothermal conditions do not hold and each face has a different lunation curve. Studies of rocks during an eclipse up
to 30 cm were also performed. The results are shown in
Fig. 39. Accuracy of the 30-cm curve is suspect because of the
isothermal assumption.

Roelof concluded that at sunset and sunrise the gross lunar
topography is probably a more important factor in temperature
anomalies, whereas the fraction of the surface covered by
centimeter rocks is too small to exhibit the effect of rock
temperatures at noon. Likewise, the enhancement in brightness
temperature during an eclipse is estimated to be not large
($\sim 15°K$) on the basis of the rock-size distribution deduced
from the observations of Surveyor I and III. Thus, small rocks
are ruled out as the cause of the larger hot spots ($\Delta T = 50°K$)
during an eclipse. This, however, may not be the case if the
rock distributions near hot spots differ markedly from the Surveyor sites. Furthermore, Roelof states that local thermal

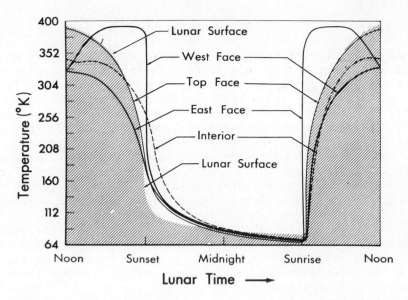

Fig. 38   Lunar temperature curves for the faces of a
          1-m basaltic cube on the lunar equator after
          Roelof.[42]

measurements by a spacecraft could be influenced by centimeter
rocks. For measurements on the lunar surface the distribution
as well as location of rocks larger than 10 cm must be known,
since the energy radiated is a sensitive function of the distribution.

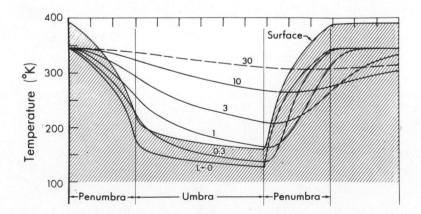

Fig. 39   Eclipse temperature curve for centimeter-size
          rocks at the apparent center of the lunar disk
          after Roelof.[42]   The shaded area represents
          the typical lunar surface.

V. Conclusions

Many characteristics of the "infrared moon" are known and
have been described in this paper. The brightness temperature
of the surface during the lunation is easily determined from
the surface albedo and a knowledge of the insolation. The directional properties of thermal emission, however, were not
predicted and required experimental measurements for their determination. These directional properties were important for
the interpretation of the Surveyor results. Recent theoretical
work by Winter[43] has shown that a certain distribution of craters with a depth-to-diameter ratio of 1:2 will reproduce the
general form of directional emission from the lunar surface.

The eclipse cooling of the average lunar surface has been
modeled by various investigators. The particulate model developed by Winter and Saari[12] has been the most successful.
In order to explain the hot spots, a distribution of rocks
similar to those found on Surveyor VII was used by Winter[43]
successfully to explain the slower cooling rates characteristic
of the eclipse hot spots.

Several questions remain concerning the thermal data. For
example, many of the hot spots cannot be identified with topo-
graphical features. In some cases, these hot spots are also
radar enhancements. Another question to be answered is the
cause of the thermal enhancements during an eclipse in the
maria or parts of maria. Outstanding in this respect is Mare
Humorum, which was found to be about 10° warmer than its environs. Mare Humorum, therefore, should be considered as a
site for further high-resolution photography, a target for orbital remote sensing from lunar orbit, and a landing site for
further exploration.

Many of these eclipse hot spots are observed by earth-based
telescopes to cool more slowly during the lunar nighttime.
Some important problems to be considered are what are the sunrise temperatures in these regions, are there any regions that
cool more slowly during the nighttime which were not observed
during the eclipse, what is the temperature distribution within the hot spot, and how large are those anomalies? In this
regard, Allen and Ney[44] have recently reported the first of
their findings on the possible nighttime temperature distribution in hot spots. They found Tycho, for instances. to have a
color temperature of 200°K, which indicates that the thermal
anomaly is probably a combination of hot regions and cold regions, with the hot regions made up of 2 to 10% of the total
area.

The exploration of the moon is continuing using earth-based telescopes, from lunar orbit, and on the lunar surface. Many of the remaining questions concerning the thermal properties of the lunar surface may be answered as these new data become available. It will not be surprising, however, if these new data also pose yet new questions to be answered.

(A slightly different version of this paper has been published in the Journal of Spacecraft and Rockets, Vol. 7, No. 4, April 1970, pp. 385-397.)

References

[1] Rosse, L., "On the Radiation of Heat from the Moon," *Proceedings of the Royal Society of London*, Vol. 27, 1869, pp. 436-441.

[2] Pettit, E. and Nicholson, S. P., "Lunar Radiation and Temperatures," *Astrophysical Journal*, Vol. 71, No. 2, Feb. 1930, pp. 102-135.

[3] Zel'tser, M. S., "The Temperature of the Lunar Surface," *The Moon - A Russian View*, edited by A. V. Markov, University of Chicago, 1960, pp. 175-203.

[4] Sinton, W. M., "Temperature on the Lunar Surface," *Physics and Astronomy of the Moon*, edited by Z. Kopal, Chap. 11, Academic, New York, 1962, pp. 407-428.

[5] Shorthill, R. W., "Measurements of Lunar Temperature Variations During an Eclipse and Throughout a Lunation," *Proceedings of the Conference on Lunar Exploration, Bulletin*, Virginia Polytechnic Institute, Engineering Experiment Station Series 152, Part A, Vol. 56, No. 7, May 1963, Paper VI.

[6] Wesselink, A. J., "Heat Conductivity and Nature of the Lunar Surface Material," *Bulletin of the Astronomical Institute, Netherlands*, Vol. 10, No. 390, April 1948, pp. 351-363.

[7] Jaeger, J. C. and Harper, A. F. A., "Nature of the Surface of the Moon," *Nature*, Vol. 166, Dec. 1950, p. 1026.

[8] Jaeger, J. C., "The Surface Temperature of the Moon," *Australian Journal of Physics*, Vol. 6, March 1953, pp. 10-21.

[9] Sinton, W. M., "The Moon," *Planets and Satellites*, edited by G. P. Kuiper, University of Chicago, 1961, pp. 438-441.

[10]Geoffrion, A. R., Korner, M. and Sinton, W. M., "Isothermal Contours of the Moon," *Lowell Observatory Bulletin No. 106*, Vol. V, No. 1, May 1960, pp. 1–15.

[11]Shorthill, R. W., Borough, H. C. and Conley, J. M., "Enhanced Lunar Thermal Radiation during a Lunar Eclipse," *Publication of the Astronautical Society of the Pacific*, Vol. 72, No. 429, Dec. 1960, pp. 481–485.

[12]Winter, D. F. and Saari, J. M., "A Particulate Thermophysical Model of the Lunar Soil," *Astrophysical Journal*, Vol. 156, June 1969, pp. 1135–1157.

[13]Sinton, W. M., "A Pyrometer for Planetary Temperature Measurements," *Lowell Observatory Bulletin No. 104*, Vol. IV, No. 16, Dec. 1959, pp. 260–263.

[14]Shorthill, R. W. and Saari, J. M., "Radiometric and Photometric Mapping of the Moon through a Lunation," *Annales of the New York Academy of Sciences*, Vol. 123, Article 2, June 1965, pp. 722–739.

[15]Saari, J. M. and Shorthill, R. W., "Isothermal and Isophotic Atlas of the Moon," NASA Contractors Rept. CR-855, Sept. 1967, NASA.

[16]Low, F. J., "Low Temperature Germanium Bolometer," *Journal of the Optical Society of America*, Vol. 51, No. 11, Nov. 1961, pp. 1300–1304.

[17]Hunt, G. R., Salisbury, J. W. and Vincent, K., "Lunar Eclipse: Infrared Images and an Anomaly of Possible Internal Origin," *Science*, Vol. 162, Oct. 1968, pp. 252–254.

[18]Pettit, E., "Radiation Measurements on the Eclipsed Moon," *Astrophysical Journal*, Vol. 91, May 1940, pp. 408–420.

[19]Saari, J. M. and Shorthill, R. W., "Isotherms of Crater Regions on the Illuminated and Eclipsed Moon, *Icarus*, Vol. 2, Aug. 1963, pp. 115–136.

[20]Murray, B. C. and Wildey, R. L., "Surface Temperature Variations During the Lunar Nighttime," *Astrophysical Journal*, Vol. 139, No. 2, Feb. 1964, pp. 734–750.

[21]Vitkus, G., Lucas, J. W. and Saari, J. M., "Lunar Surface Thermal Characteristics during Eclipse from Surveyors III, V, and after Sunset from Surveyor V," AIAA Paper 68-747, Los Angeles, Calif., 1968.

[22]Wildey, R. L., Murray, B. C. and Westphal, J. A., "Reconnaissance of Infrared Emission from the Lunar Nighttime Surface," *Journal of Geophysical Research*, Vol. 72, July 1967, pp. 3743-3749.

[23]Low, F. J., "Lunar Nighttime Temperatures Measured at 20 Microns," *Astrophysical Journal*, Vol. 142, No. 2, June 1965, pp. 806-808.

[24]Mendell, W. and Low, F. J., "Low-Resolution Differential Drift Scans of the Moon at 22 Microns," *Journal of Geophysical Research*, Vol. 75, No. 17, June 1970, pp. 3319-3324.

[25]Ingrao, H. C., Young, A. T. and Linsky, J. L., "A Critical Analysis of Lunar Temperature Measurements in the Infrared," *Rept. 6*, April 1965, Harvard College Observatory of Science.

[26]LaRocca, A. and Zissis, G. J., "Field Sources of Blackbody Radiation," *The Review of Scientific Instruments*, Vol. 30, No. 3, March 1959, pp. 200-201.

[27]Shorthill, R. W. and Saari, J. M., "Infrared Observations on Eclipsed Moon," *Physics and Astronomy of the Moon*, 2nd ed., edited by Z. Kopal, Chap. 9, Academic, New York (to be published).

[28]Lucas, J. W., Conel, J. E., Hagemeyer, W. A. and Saari, J. M., "Lunar Surface Thermal Characteristics from Surveyor I," *Journal of Geophysical Research*, Vol. 72, No. 2, Jan. 1967, pp. 779-789.

[29]Lucas, J. W., Garipay, R. R., Hagemeyer, W. A., Saari, J. M. Smith, J. W. and Vitkus, G., "Lunar Surface Temperatures and Thermal Characteristics: Surveyor V Science Results," *Journal of Geophysical Research*, Vol. 73, No. 22, Nov. 1968, pp. 7209-7219.

[30]Vitkus, G., Garipay, R. R., Hagemeyer, W. A., Lucas, J. W. and Saari, J. M., "Lunar Surface Temperatures and Thermal Characteristics," *TR 32-2362*, Surveyor VI Mission Report – Part II: Mission Results, Jan. 1968, pp. 109-123, Jet Propulsion Laboratory.

[31]Montgomery, C. G., Saari, J. M., Shorthill, R. W. and Six, Jr., N. F., "Directional Characteristics of Lunar Thermal Emission," *Document D1-82-0568*, Nov. 1966, Boeing Scientific Research Laboratories.

[32]Saari, J. M., "The Surface Temperature of the Antisolar Point of the Moon," *Icarus,* Vol. 3, No. 2, July 1964, pp. 161-163.

[33]Ingrao, H. C., Young, A. T. and Linsky, J. L., "A Critical Analysis of Lunar Temperature Measurements in the Infrared," *The Nature of the Lunar Surface,* edited by W. Hess, D. Menzel, and J. O'Keefe, John Hopkins, Baltimore, 1966, pp. 185-211.

[34]Hunt, G. R., Salisbury, J. W. and Vincent, R. K., "Infrared Images of the Eclipsed Moon," *Sky and Telescope,* Vol. 36, No. 4, Oct. 1968, pp. 2-4.

[35]Saari, J. M. and Shorthill, R. W., "Infrared and Visible Images of the Eclipsed Moon of December 19, 1964," *Icarus,* Vol. 5, No. 6, Nov. 1966, pp. 635-659.

[36]Shorthill, R. W. and Saari, J. M., "Nonuniform Cooling of the Eclipsed Moon: A Listing of Thirty Prominent Anomalies," *Science,* Vol. 150, No. 3693, Oct. 1965, pp. 210-212.

[37]Green, R. R., "An Analysis of the Distribution of the Major Surface Characteristics and the Thermal Anomalies Observed on the Eclipsed Moon," *Document D1-82-0775,* Jan. 1969, Boeing Scientific Research Laboratories.

[38]"Radar Studies of the Moon," *Lincoln Lab. Report,* April 1968, Massachusetts Institute of Technology.

[39]Saari, J. M., "Lunar Thermal Anomalies and Internal Heating," *Astrophysics and Space Sciences,* Vol. 4, Jan. 1969, pp. 275-284.

[40]Middlehurst, B., Burley, J. M., Moore, P. and Wilther, B. L., "Chronological Catalog of Reported Lunar Events," *TR-277,* Vol. 4, No. 63, July 1968, NASA.

[41]Watson, K., "I. Thermal Conductivity Measurements of Selected Silicate Powders in Vacuum from 150° - 350°K; II: An Interpretation of the Moon's Eclipse and Lunation Cooling as Observed through the Earth's Atmosphere from 8-14 Microns," unpublished Ph.D. Thesis, California Institute of Technology, 1964.

[42]Roelof, E. C., "Thermal Behavior of Rocks on the Lunar Surface," *Icarus,* Vol. 8, No. 1, Jan. 1968, pp. 138-159.

[43]Winter, D. F., "The Infrared Moon: Data, Interpretations, and Implications," *Document D1-82-0090*, Sept. 1969, Boeing Scientific Research Laboratories.

[44]Allen, D. A. and Ney, E. P., "Lunar Thermal Anomalies: Infrared Observations," *Science*, Vol. 164, April 1969, pp. 419-421.

MICROWAVE EMISSION FROM THE MOON

Duane O. Muhleman*

California Institute of Technology, Pasadena, Calif.

Introduction

Measurements of blackbody thermal emission from the moon in the infrared spectral region yield much information concerning the upper few centimeters of the lunar crustal material. (This chapter is intended to be tutorial in nature rather than an exhaustive review of the literature. Consequently, many interesting but perhaps more speculative developments are not discussed. The reader will be referred to more extensive reviews cited in the text.) Obviously, one can study the structure of this material at greater depths from measurements of thermal emission at longer wavelengths, i.e., the radio region of the electromagnetic spectrum. Significant measurements have been obtained over lunation periods and during lunar eclipses for wavelengths ranging from 1 mm to nearly 1 m. The emission mechanism is exclusively thermal in origin, and details of the emission such as polarization, variations with wavelength, lunar phase, etc., are controlled by the thermal and electrical parameters of the lunar surface material to a depth of several meters (at the longest wavelengths).

Investigations of the radio emission from the moon are closely related to the extensive studies of radar reflections from the lunar surface. The latter subject is developed in detail in Chap. 1c. In this chapter, we shall consider only aspects of this work which are directly important in interpreting the radio emission observations.

If we consider the lunar surface to be a smooth semi-infinite slab that is homogeneous in terms of at least electrical properties, we can compute the equivalent blackbody brightness temperature that one would observe at a frequency ν

Supported in part by NASA grant NGR-005-002-114.
*Professor of Planetary Sciences, Division of Geological and Planetary Sciences, and the Owens Valley Radio Observatory.

and at an angle of incidence θ_o to the surface if the
temperature distribution with depth, $T(x)$, is known. The
resulting radiative transfer problem was solved by Piddington
and Minnett[20] in the form

$$T_B(\nu,p) = \left[1 - R_p(\nu,\theta_o)\right] \int_0^\infty T(x) \exp\left[-k_\nu x/\cos\theta_i\right] k_\nu \, dx/\cos\theta_i \tag{1}$$

[the integral being the Laplace transform of $T(x)$], where
k_ν = the power absorption coefficient at frequency ν; θ_i = the
angle of incidence of pencil radiation just below the surface,
which is related to the incidence angle of observation θ_o by
Snell's law at the surface; p = the state of polarization,
either parallel or perpendicular to the plane of incidence; and
$R_p(\nu,\theta_o)$ = the Fresnel reflection coefficient of polarization
p for radiation emerging at angle θ_o. The Fresnel reflection
coefficients for the E vector parallel to the incidence plane
$R_\parallel$ and perpendicular to the incidence plane $R_\perp$ are given by

$$R_\parallel(\nu,\theta_o) = \left[\frac{\epsilon \cos\theta_o - \sqrt{\epsilon - \sin^2\theta_o}}{\epsilon \cos\theta_o + \sqrt{\epsilon - \sin^2\theta_o}}\right]^2 \tag{2}$$

$$R_\perp(\nu,\theta_o) = \left[\frac{\cos\theta_o - \sqrt{\epsilon - \sin^2\theta_o}}{\cos\theta_o + \sqrt{\epsilon - \sin^2\theta_o}}\right]^2 \tag{3}$$

where ϵ is the complex dielectric constant. The quantity
actually measured in radio astronomy is the flux density (in
w/m^2-Hz). For the temperature range of the moon, the Planck
radiation function at wavelengths greater than about 1 mm can
be replaced by the Rayleigh-Jeans approximation, and the flux
density in a pencil of radiation in a solid angle Ω can be
written as

$$F = (2kT/\lambda^2)\Omega \tag{4}$$

(λ is the wavelength of the radiation in the material)
where k = Boltzmann's constant. Equation (4) was used in the
derivation of Eq. (1) and is used to express the measured flux
densities in terms of equivalent blackbody brightness temperatures $T_B(\nu,p)$.

Equations (1-3) contain the complex dielectric constant of
the lunar surface material, where the real part ϵ primarily
controls the reflection phenomena at the very surface and the
imaginary part is expressed in terms of the (power) absorption
coefficient k_ν, which supplies the opacity within the material.
The absorption coefficient is strongly wavelength-dependent
for dielectric materials (ϵ is not), and it is given to sufficient accuracy for lunar conditions by

$$k_\nu = k_o \nu \tag{5}$$

over the frequency range of interest. Troitsky et al.[23] have
pointed out that the constant k_o may be temperature-dependent
in the lunar application. It can be seen from Eqs. (1) and
(5) that, with decreasing frequencies, k_ν has the effect of
making the temperature distributions at greater depths in the
material more important at a given incidence angle.

Equation (1) has been used in all interpretations of the
lunar observations known to us. Consequently, it is important
to consider carefully all of the assumptions involved. First
of all, $T_B(\nu,p)$ is the equivalent temperature that would
obtain with an infinitely narrow antenna beam with polarization p. All real observations involve an antenna beam pattern
of finite width (often larger than the lunar disk itself)
which necessarily averages the emission over a range of incidence angles θ_o and, consequently, over a range of polarization planes relative to the local spherical surface of the
moon. The appropriate temperature distribution can no longer
be considered just a function of depth in the integration over
the beam pattern. The most serious assumption leading to Eq.
(1) is the homogeneity of the surface material in the upper
few meters of the lunar soil. Both ϵ and k_ν are fairly strong
functions of the density of a specified material. Variations
in radar cross section of the moon with wavelength from radar
measurements strongly suggest a density increase with depth.
A rigorous treatment of the problem for a given density profile in x would require a consideration of ray-path bending
in the material due to refraction. However, the accuracy of
the radio observations probably does not warrant these
detailed considerations. The effects of surface roughness on
the emission, particularly near grazing incidence, are poorly
understood.

Linear Theory

The radio observations of the moon primarily show a sinus-
oidal variation about a mean brightness with the lunation

period. The amplitude and phase of this variation
systematically depend on the wavelength, i.e., the amplitude
of the variation increases with increasing frequency, and the
phase approaches that of the solar insolation. No significant
variation can be seen at wavelengths longer than about 5 cm.
The observations can be fully exploited only after one has
computed a detailed thermophysical model for the moon, i.e.,
the temperature distribution with longitude and latitude to
depths of several meters. Under certain assumptions, the
entire problem can be solved analytically as shown by
Piddington and Minnett (1949)[20] and Troitsky[21]. Troitsky in
particular has greatly developed the method in a distinguished
series of papers. (See Troitsky and Tikhonova[24] for complete
references and a summary of results.) In this approach, which
we call the <u>linear theory</u>, we begin by assuming that the temperature at a given longitude and latitude point on the
<u>surface</u> of the moon can be expanded in a Fourier series in
time

$$T(0,t) = T_o + T_1 \cos (\omega t - \Phi_1) + T_2 \cos (2\omega t - \Phi_2) + \ldots \quad (6)$$

where ω is the angular rotational rate of the moon with
respect to the sun and Φ_n is the phase shift of the nth harmonic term relative to the solar insolation. If we assume
that the moon is a semi-infinite homogeneous slab with <u>constant</u>
thermal properties, the equilibrium solution for the temperature at depth x has been shown to be[2]

$$T(x,t) = T_o + T_1 e^{-\beta_1 x} \cos (\omega t - \Phi_1 - \beta_1 x) +$$

$$T_2 e^{-\beta_2 x} \cos (2\omega t - \Phi_2 - \beta_2 x) + \ldots \quad (7)$$

where β_n is the "thermal absorption coefficient" for the nth
harmonic and is given by

$$\beta_n = \left[\frac{n}{2} \frac{\rho c_p}{k} \right]^{1/2} \quad (8)$$

where ρ is the density, c_p the specific heat, and k the thermal conductivity of the material, all of which are constants
(with depth and temperature) in this development.

The physical interpretation of Eq. (7) is straightforward.
The temperature at any depth x is the sum of a constant term
T_o and a series of periodic terms arising from harmonic temperature "waves" that propagate into the material with

amplitudes that are attenuated with depth by virtue of the $e^{-\beta_n x}$. The phases of the temperature waves are increasingly retarded with depth by the same factor in the arguments of the cosine terms. The temperature distribution given by Eq. (7) can be utilized in Eq. (1) to compute the radio brightness temperatures at any wavelength for a given set of electrical constants ϵ and k_ν. The angles θ_o and θ_i in Eq. (1) are, of course, determined by the line of sight from the earth to the particular point on the lunar surface for which the temperature distribution with depth is given by Eq. (7). The integral in Eq. (1) becomes an infinite series of integrals, the first two terms of which are the most important. For the constant term in Eq. (7), we get

$$T_{B_o}(\nu,p) = \left[1 - R_p(\nu,\theta_o)\right] \int_0^\infty T_o \exp\left[-k_\nu x/\cos\theta_i\right] k_\nu dx/\cos\theta_i$$

or $\hspace{11cm}$ (9)

$$T_{B_o}(\nu,p) = \left[1 - R_p(\nu,\theta_o)\right] T_o$$

that is, essentially a constant for a given θ_o since R_p is practically frequency–independent. The general nth term becomes (at time t)

$$T_{B_n}(\nu,p) = \left[1 - R_p(\nu,\theta_o)\right] T_n \int_0^\infty \exp\left[-(k_\nu/\cos\theta_i + \beta_n)x\right] \times$$

$$\cos\left[n\omega t - \Phi_n - \beta_n x\right] k_\nu dx/\cos\theta_i \hspace{2cm} (10)$$

We define the fundamental parameter $\delta_n(\theta_i)$ as

$$\delta_n(\theta_i) = \left(\frac{\beta_n}{k_\nu}\right)\cos\theta_i = \left(\frac{n^{1/2}\beta_1}{k_\nu}\right)\cos\theta_i \hspace{2cm} (11)$$

The quantity β_1/k_ν is the ratio of the thermal (temperature) absorption coefficient to the radio absorption coefficient, or, introducing the thermal absorption length ℓ_T and the radio absorption length ℓ_R, we see that $\delta_1(\theta_i)$ at normal incidence is the fundamental ratio

$$\delta_1(\nu) = \beta_1/k_\nu = \ell_{R,\nu}/\ell_T \tag{12}$$

We shall see shortly that the quantities $\delta_1(\nu)$ are directly measured by the radio observations during a lunation. The integrals, Eq. (10), are elementary and, after a rather long manipulation, yield the remarkable result (at time t at a specified point on the lunar surface)

$$T_{B_n}(\nu,p) = \left[1 - R_p(\nu,\theta_0)\right] \left\{ \frac{T_n \cos\left[n\omega t - \Phi_n - \Psi_n(\theta_i)\right]}{\sqrt{1 + 2\delta_n(\theta_i) + 2\delta_n^2(\theta_i)}} \right\} \tag{13}$$

where the phase factor of the brightness temperature $\Psi_n(\theta_i)$

$$\Psi_n(\theta_i) = \arctan\left[\delta_n(\theta_i)/(1 + \delta_n(\theta_i))\right] \tag{14}$$

Finally, the complete solution of the thermophysical problem in the linear theory is

$$T_B(\nu,p) = \left[1 - R_p(\nu,\theta_0)\right] \left\{ T_0 + \sum_{n=1}^{\infty} T_{B_n}(\nu,p) \right\} \tag{15}$$

In all of our equations, we have carried both angles θ_0 and θ_i. They are, of course, related by Snell's law applied at the vacuum-surface interface:

$$\sin\theta_0 = \sqrt{\epsilon}\,\sin\theta_i \tag{16}$$

where Maxwell's relation is used to write the index of refraction as the square root of the (real) dielectric constant. Since the electrical conductivity of the lunar surface material is small, the use of the real part of the dielectric constant in Eq. (16) and in the Fresnel reflection coefficients, Eqs. (2) and (3), will not cause significant errors in the interpretation of the radio observations.

Equations (13) through (16) formally complete the linear theory. However, it is useful here to recapitulate the development in order to emphasize the simple physical ideas. We started with the knowledge of the <u>surface</u> temperature as a function of time at a specified point on the spherical surface of the moon. Any such point can easily be related to an observer on the earth as a function of time from the motion of the earth-moon-sun system. The temperature variation was then expanded in an infinite Fourier series, Eq. (6), which is always possible mathematically. The lunar surface was

then replaced by a semi-infinite homogeneous slab at the local
point. This assumption can cause no errors, since the ratio of
the thermal skin depth to the radius of the moon is infinites-
imal. The resulting temperature distribution with depth below
the local point was then given by Eq. (7) under the assumption
that the parameters k, ρ, and c_p are independent both of depth
and of temperature. We shall see below that these assumptions
are serious and that this theory is inadequate to explain some
of the details of the observations, although it does accurately predict the major features of the observations. It
should be realized that the Fourier expansion approach is not
limited to the conditions of lunations. Any surface temperature variation can be treated in this way, including eclipse
circumstances, although its usefulness seems to be limited to
lunation conditions. The temperature distribution with depth
retains the harmonic terms of the surface variation. However,
each higher harmonic is increasingly attenuated by a factor of
n times the thermal absorption coefficient. Furthermore, each
higher harmonic is increasingly delayed in phase relative to
the surface variation by the same factor. Only at the shortest
wavelengths can harmonics higher than the first be seen in the
radio emission, as we shall see below. The thermal (temperature) absorption coefficient is related to the thermal parameters and the rotational rate of the moon by Eq. (8).

The temperature distribution with depth, $T(x,t)$, was then
utilized in the formal solution of the radiative transfer
problem, Eq. (1), which was derived under the assumptions that
electrical parameters are independent of depth and temperature
and that the surface is smooth. Departures from these conditions obviously occur on the moon, but they are probably not
as serious as the assumptions concerning the thermal parameters. At the time of this writing, the effects of variations
in the electrical parameters have not been investigated for the
complete lunar thermophysical problem. However, these questions will be considered for the determination of ϵ later in
this chapter. The equivalent brightness temperature at a
specified frequency (contained in k_ν) at incidence angle θ_o was
given by Eqs. (13-15), the constant plus first harmonic terms
being

$$T_B(t) = \left[1 - R_p(\theta_o)\right]\left\{T_o + \frac{T_1}{\sqrt{1 + 2\delta_1(\theta_i) + 2\delta_1^2(\theta_i)}} \times \right.$$

$$\left. \cos\left[\omega t - \Phi_1 - \Psi_1(\theta_i)\right]\right\} \tag{17}$$

where $\delta_1(\theta_i)$ was given by Eq. (11) and $\Psi_1(\theta_i)$ by Eq. (14).
Since k_ν is proportional to ν, δ increases with decreasing
frequency. The brightness temperature consists of a constant
term independent of frequency plus an harmonic term whose amplitude has decreased from the surface values by the division
by the factor $[1 + 2\delta + 2\delta^2]^{1/2}$, which increases with increasing wavelength. At long wavelengths, T_B approaches the mean
temperature T_o times the emissivity $(1 - R)$. Furthermore, the
phase of T_B increases relative to that of the insolation as
the wavelength increases, reaching a maximum of 45° as δ becomes large, as can be seen from Eq. (14).

Thus far, we have considered the case for observations with
an infinitely narrow antenna beam. For a finite beam, Eq. (15)
has to be integrated over the beam pattern, which means inte-
grating over a range of θ_o and θ_i as well as over the phase
curve, i.e., a range of "time" t. These calculations are complex and will not be considered here (see, e.g., Kaydonovsky
et al.[11]). Lunar radio observations can be corrected for the
antenna patterns and referred to normal incidence at the subearth point with some loss of information. It is these observations that we discuss below. Observations so corrected are
easy to interpret, since $\theta_o = \theta_i = 0$ and the Fresnel reflection coefficients become the (polarization-independent) normal
power reflection coefficient, which is determined nearly
directly by radar cross-section measurements.

Determination of δ from the Observations

A list of radio brightness temperatures over lunations
expressed as a mean T_m, amplitude of the first harmonic T_v,
and phase delay of the first harmonic Ψ are shown in Table 1
(compiled by Hagfors[8,9]). We have plotted the observed ratios
T_m/T_v against the wavelength in Fig. 1. The theoretical value
of this ratio from Eq. (17) is (for normal incidence)

$$T_m/T_v = T_o/T_1\left[1 + 2\delta_1 + 2\delta_1^2\right]^{1/2} \tag{18}$$

Following Troitsky[22], we adopt the working hypothesis that δ_1
is proportional to the wavelength over the wavelength range
of interest. With this hypothesis, based on the available
observations at that time, Troitsky adopted the relationship
$\delta = 2\lambda$, where λ is in centimeters. The curve for the relationship is shown in Fig. 1. This curve, as well as the curve for
$\delta = 2.5\lambda$, was forced to agree with the observations near
0.1 cm, since the ratio T_o/T_1 is not known. The ratio apparently should be interpreted as the ratio of the mean <u>surface</u>
temperature to the amplitude of the first harmonic of the

Table 1 List of results of measurements of lunation variation
of thermal emission from the moon

Wavelength, cm	T_o, °K	T_i, °K	ϕ_i, deg	Half-power beam width	References (cited by Hagfors[8],[9])
0.10a	229	115	18	3.9'	Low and Davidson [1965]
0.13a	216	120	16	10'	Fedoseev [1963]
0.15b	265	145	...	5'	Sinton [1955]
0.18a	240	115	14	6'	Naumov [1963]
0.32a	210	65	10	9'	Tolbert and Coates [1963]
0.33a	196	70	27	2.9'	Gary et al. [1965]
0.40b	230	73	24	25'	Kislayakov [1961]
0.40a	228	85	27	1.6'	Kislayakov and Salomonvich [1963]
0.40b	204	56	23	36'	Kislayakov and Plechkov [1964]
0.80a	197	32	40	18'	Salomonovich [1958]
0.80a	211	40	30	2'	Salomonovich and Losovskii [1963]
0.86a	225	45	40	6'	Gibson [1958]
1.18a	220	29	48	3.5'	Moran [1965]
1.25a	215	35	45	23'	Piddington and Minnett [1949]
1.37a	220	24	43	4.0'	Moran [1965]
1.6b	208	37	30	44'	Kamenskaya et al. [1962]
1.63b	224	36	34	26'	Zelinskaya et al. [1959]
1.63b	207	32	10	44'	Dimitrenko et al. [1964]
2.0a	190	20	40	4'	Salomonovich and Koschenko [1961]
3.15a	195	12	44	9'	Mayer [1961]
3.2a	223	17	45	6'	Koschenko et al. [1961]
3.2b	216	16	15	40'	Bondar et al. [1962]
9.4b	220	5.5	5	2°20'	Medd and Broten [1961]
9.6b	230	...	...	19'	Koschenko et al. [1961]
11b	214	...	...	17'	Mezger and Straussl [1959]
20.8b	205	5	49	36'	Waak [1961]
21.0b	232	...	...	10'	Heiles and Drake [1963]

Table 1 (continued)

Wavelength, cm	T_o, °K	T_i, °K	ϕ_i, deg	Half-power beam width	References (cited by Hagfors[8,9])
21.0b	250	5	...	35'	Mezger and Strassl [1959]
21.0b	230	...	...	90'	Razin and Fedorev [1963]
22	270	...	...	15'	Davis and Jenisson [1960]
25.0b	226	...	...	1°17'	Alekseev et al. [1967]
30.2b	227	...	...	1°18'	Alekseev et al. [1967]
31.25b	227	...	...	...	Troitsky et al. [1967]
32.3b	233	...	...	3°00'	Razin and Fedorev [1963]
35.2b	223	...	...	...	Troitsky et al. [1967]
40b	224	...	...	...	Troitsky et al. [1967]
54b	218.5	...	...	...	Troitsky et al. [1967]
60.24b	216.5	...	...	...	Troitsky et al. [1967]
70.16b	217	...	...	1°30'	Krotikov et al. [1964]

[a]Brightness measured at center of disk.

[b]Brightness averaged over disk.

surface temperature. The true surface temperature determined from infrared observations does, of course, contain higher harmonics. The curves of Fig. 1 yield $T_o/T_1 \approx 1.5$. If the mean surface temperature is 240°K, then the maximum and minimum temperatures of the first harmonic would be about 400 and 80°K, respectively. It can be seen from the scatter of the data in Fig. 1 that, although the linear theory agrees rather well with the observations, the hypothesis that δ_1 is proportional to the wavelength is only weakly supported. Consistent with this, we shall adopt the result

$$\delta_1 = (2.4 \pm 0.5)\lambda \qquad (19)$$

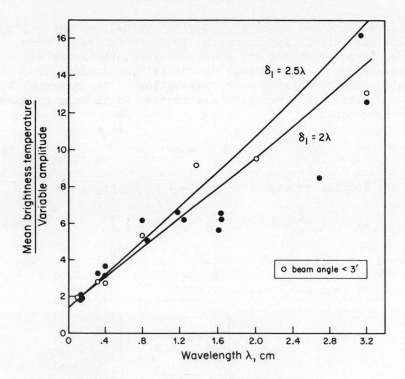

Fig. 1   Ratios of the mean temperatures to the variable temperature amplitudes during a lunation. Circle points are measurements with
very narrow beamed antenna systems.

in further discussions. The temperature phase data of Table 1
are also consistent with this result.

Interpretation of δ_1

The value δ_1 which was obtained from the observations is
related to the physical parameters of the lunar material
through Eq. (12), i.e., it is the ratio of the electrical
length at frequency ν, $\ell_{R,\nu}$, to the thermal length ℓ_T. It is,
of course, of considerable interest to separate the physical
parameters from the empirical parameter δ_1. This cannot be
done from the radio observations alone. A considerable body
of information is contained in the infrared and radar measurements of the lunar surface. These data, combined with laboratory measurements of likely lunar materials (see Chap. 3a) and
recently with returned lunar samples (see Chap. 3b), allow for
a nearly unique separation of the physical parameters.

Before we consider specific numerical values, we shall transform the electrical coefficient k_ν into a more useful form.
Since k_ν is not zero for the lunar material, the material
possesses a nonzero electrical conductivity. A plane mono-
chromatic electromagnetic wave propagating in the material will
have an electric vector $\overline{E}$ harmonic in time which is a solution
of the wave equation

$$\nabla^2 \overline{E} + k_c^2 \overline{E} = 0 \tag{20}$$

where the complex propagation constant k_c is given by

$$k_c^2 = (\omega^2 \mu/c^2)\left[\epsilon + i\ 4\pi\sigma/\omega\right] \tag{21}$$

($\omega = 2\pi\nu$, c = light speed in vacuum, σ = specific electrical
conductivity, and μ = magnetic permeability). Writing k_c in
terms of its real and imaginary parts,

$$\mathrm{Re}(k_c) = \frac{\omega}{c}\sqrt{\frac{\mu\epsilon}{2}}\left\{\sqrt{1 + \frac{4\sigma^2}{\epsilon^2\nu^2}} + 1\right\}^{1/2} \tag{22}$$

$$\mathrm{Im}(k_c) = \frac{\omega}{c}\sqrt{\frac{\mu\epsilon}{2}}\left\{\sqrt{1 + \frac{4\sigma^2}{\epsilon^2\nu^2}} - 1\right\}^{1/2} \tag{23}$$

The electric vector $\overline{E}$ of the plane wave will be absorbed as it
propagates through the material by virtue of the imaginary
part of k_c. The absorption of the intensity or power in the
wave is given by the square of $\overline{E}$ absorption, or

$$\overline{E}^2 \propto \overline{E}_o^2\ e^{-2\mathrm{Im}(k_c)x}$$

for propagation in the x direction. Thus, the power absorption coefficient is twice $\mathrm{Im}(k_c)$:

$$k_\nu = 2\mathrm{Im}(k_c) = \frac{\omega}{c}\sqrt{2\mu\epsilon}\left\{\sqrt{1 + \frac{4\sigma^2}{\epsilon^2\nu^2}} - 1\right\} \tag{24}$$

It has become conventional in the literature on lunar radio
emission to express the electrical absorption in terms of the
loss tangent, which is defined as

$$\tan\Delta = 2\sigma/\epsilon\nu \tag{25}$$

In the microwave region of the spectrum, the loss tangents for almost all materials occurring on the earth's surface which are free of liquid water are considerably less than unity, i.e., the materials possess low electrical conductivities at high frequencies. This circumstance undoubtedly is also true for the lunar surface material. With this mild assumption, the absorption coefficient becomes

$$k_\nu = (\omega/c)\sqrt{\mu\epsilon}\,\tan\Delta$$

Except for magnetic materials, which are probably not important on the lunar surface, we can set $\mu = 1$, and the power absorption coefficient becomes

$$k_\nu = (2\pi\nu/c)\sqrt{\epsilon}\,\tan\Delta \qquad (26)$$

or, expressed as the radio absorption length (skin depth),

$$\ell_{R,\nu} = \lambda/(2\pi\sqrt{\epsilon}\,\tan\Delta) \qquad (27)$$

Now, from Eqs. (12) and (8),

$$\delta_1 = \frac{\ell_{R,\nu}}{\ell_T} = \left\{ \left(\frac{\omega\rho c_p}{2k}\right)^{1/2} \left(2\pi\sqrt{\epsilon}\,\tan\Delta\right)^{-1}\lambda \right\} \qquad (28)$$

Our empirical result, $\delta_1 = (2.4 \pm 0.5)\lambda$, strongly suggests that the loss tangent of the material comprising the upper meter of the lunar surface is frequency independent, which, in turn, by virtue of Eq. (25), suggests that the electrical conductivity σ depends linearly on the frequency. These results conform well to laboratory measurements on dielectric materials. Actually, we anticipated this result with Eq. (5). Equation (28) is the desired factorization of δ_1 into the fundamental physical parameters. The basic result of the linear theory can be written as

$$\sqrt{\frac{\omega\rho c_p}{2k}}\,(2\pi\sqrt{\epsilon}\,\tan\Delta)^{-1} = 2.4 \pm 0.5 \qquad (29)$$

Many investigators have attempted to utilize this result with a combination of infrared and radar data to separate the numerical values of the physical parameters (e.g., see Troitsky[22] for a careful analysis of these matters and for complete references). Since we believe that the linear theory is not adequate to interpret the observations fully, we shall

not labor over these results but rather shall adopt a
somewhat arbitary set of numerical values as an illustration.
A more complete discussion will be deferred to later in this
chapter, when we consider the variation of the thermal conductivity and the specific heat with temperature. Unfortunately,
a careful analysis of the manifold assumptions concerning the
electrical parameters does not exist. It appears that the
radio observations are not sufficiently accurate at this time
to warrant such detail.

A Consistent Set of Constant Parameters

The parameter that is most in doubt is $\tan \Delta$, which we shall
adjust to yield the empirical δ_1. Our best estimate of
constant parameter values is (illustration only) $\epsilon = 2.0$,
$\rho = 1.0$ g/cm^3, $c_p = 0.6$ joules/g-$^\circ$K, and $k = 2.92 \times 10^{-5}$
joules/cm-sec-$^\circ$K, where the thermal conductivity was computed
from the listed values of ρ and c_p and a rather arbitrary
value of the thermal parameter $\gamma = (k\rho c_p)^{-1/2} = 1000$. This
parameter set is probably valid for the upper meter of the
lunar soil averaged over the lunar disk facing the earth at a
mean temperature of about 240°K. From Eq. (29), we find

$$\tan \Delta = 0.0074 \pm 0.0022 \tag{30}$$

Winter and Saari[27] have interpreted laboratory data on powders
of basaltic material and of pumice in the form

$$\tan \Delta \approx 0.006 \; [1 + 0.5 \; e - (400 - T)/50] \tag{31}$$

The agreement of our result, Eq. (30), with the laboratory
measurements is excellent. From Eq. (30), we compute the
electrical skin depth (distance over which the power will be
attenuated to e^{-1}) as $\ell_{R,\nu} = 15.2\lambda$, and the temperature depth
(distance over which the temperature of the first harmonic will
be attenuated to e^{-1}) is $\ell_T = 6.3$ cm. It is obvious from
these results why no significant variation in the microwave
brightness temperatures can be seen beyond a wavelength of
about 3 cm.

With this, we conclude our study of the linear theory. A
great deal of information concerning the moon has been obtained
from the simplified theory. A considerable body of work has
been performed beyond that discussed, including investigations
of two-layer models, which appear to be required to explain
the infrared measurements better. We believe that this work
is primarily of historical interest and that the problem is
better treated with numerical techniques in which important
nonlinear effects in the parameters can be included. Before

we review these developments, we shall carefully consider the
variations to be expected of the dielectric constant and the
thermal properties.

Physical Parameters

Thus far, we have considered a highly simplified model of
the moon which is completely homogeneous, with constant parameters independent of both depth and temperature. This model
is not sufficient to explain the details of the radio and
infrared emission, and it ignores a considerable fraction of
our current knowledge of the lunar surface. The basic parameters of lunar thermophysics consist of the dielectric constant, electrical conductivity (or loss tangent), thermal
conductivity, specific heat, and the material density. In
particular, the density certainly varies with depth and with
location on the lunar surface, variations which will affect
all of the remaining parameters. These parameters also vary
with ambient temperature, although the variations of the
electrical parameters are probably negligible within the
temperature range of interest. A considerable body of information is available concerning the parameters from diverse
lunar observation and laboratory measurements. In this section, we briefly discuss these results; a more thorough discussion of the thermal parameters can be found elsewhere in
this volume. We shall begin with the dielectric constant and
the loss tangent.

The dielectric constant ϵ has been obtained by two methods:
1) the measurement of radar cross sections of the moon over a
range of wavelengths from 0.86 cm to 2 m, and 2) measurements
of the polarization of the radio emission from the moon over
a range of wavelengths from 0.86 to 21 cm. The methods
differ in a number of significant ways. The radar method
yields estimates of ϵ near the subearth point averaged over
depths of a few wavelengths, whereas the polarization measurements are sensitive to the material from the Brewster angle
(~55 deg from the subearth point) to the limb of the moon over
a smaller range of depths. Radar cross-section measurements
are very sensitive to the surface roughness (i.e., the radar
backscatter law), whereas surface roughness causes a relatively
small but significant depolarization of the radio emission.
Finally, radar measurements are easily obtained over a large
range of wavelengths, whereas polarization measurements must
be performed with narrow-beam systems and are technically more
difficult.

The radar cross section is defined as the ratio of the measured echo power (for continuous-wave illumination or long pulses) to the power that would be reflected from an equivalent smooth and perfectly conducting sphere. It can be written as

$$S_\lambda = g_\lambda R \tag{32}$$

where R is the power reflection coefficient at normal incidence and g_λ is the backscatter "gain" of the surface or the directivity of the surface. The directivity arises from surface roughness and, in general, is wavelength-dependent. It is formally defined as the ratio of the power scattered back to the radar per unit solid angle to the power scattered into the average unit solid angle. The directivity has never been measured for the lunar surface, since such measurements would require bistatic experiments over the complete range of angles. Backscattering theories that describe lunar radar echoes have been developed by Hagfors[6] and Muhleman[17], and these authors have obtained similar expressions for directivities. Using measured pulse responses of the moon at wavelengths 3.8, 23, and 68 cm and the theoretical results just cited, we have estimated an empirical interpolation expression for the directivities as a function of wavelength

$$g_\lambda \approx 1 + \lambda^{-0.56} \quad (1 \text{ cm} \leq \lambda) \tag{33}$$

where λ is expressed in centimeters. Radar cross-section data from Hagfors[8], are shown in Fig. 2, where we have added a smooth line with error limits. The observations are very difficult for long wavelengths and are probably not reliable beyond 1 m. Using Eq. (33) for g_λ, and either Eq. (2) or (3) for the relationship between $R(\theta_o)$ at $\theta_o = 0$ and ϵ, we obtain the estimates for ϵ shown in Fig. 3, which increase from $\epsilon = 2$ at short wavelengths to $\epsilon \sim 5$ (with a large uncertainty) at the longest wavelengths. This result can be interpreted in terms of the variation in density with depth into the lunar surface. The lunar basalts in a fully consolidated form have a dielectric constant in the range of $\epsilon_s = 5$ to 7. If we take the density of this material to be $\rho_s = 3.0$ g/cm^3 on the basis of Apollo returned samples, the dielectric constant for a porous state of the basalt can be computed from the empirical expression

$$\epsilon = \epsilon_s \left[1 - \frac{3p}{[(2\epsilon_s + 1)/(\epsilon_s - 1)] + p} \right] \tag{34}$$

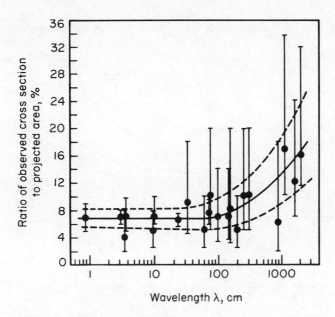

Fig. 2   Radar cross-sectional measurements
as a function of wavelength. The
smooth line represents moving
averages and the dashed lines,
approximate error bounds.

where $p = 1 - \rho/\rho_s$ (Troitsky[22]; see also Campbell and Ulrichs[1]
for a more recent discussion). The corresponding densities,
determined from the radar cross-section data, can be read from
the right-hand scale in Fig. 3 (where $\epsilon_s = 6$). It would
indeed be an important contribution if we could relate the
density-vs-wavelength curve in terms of actual depth into the
lunar surface. Although several investigators have attempted
this, we believe that their procedures are quite ambiguous.

Estimates of the dielectric constant for four selected sites
on the lunar surface have been obtained from the Surveyor
radar altimeter data by Muhleman[18]. These measurements were
made at a wavelength of 2.3 cm and refer to regions smaller
than a square kilometer. The results are shown in Table 2,
where we have computed the density from Eq. (34) with $\epsilon_s = 6$
and $\rho_s = 3$ g/cm^3. The uncertainties quoted in the table are
almost completely due to the uncertainty in determining the
directivity factors. The first three Surveyor sites are mare
regions, and the Surveyor VII site is on the Tycho blanket
region. The latter region displays a much higher dielectric
constant and density, which <u>may</u> be typical of upland regions.

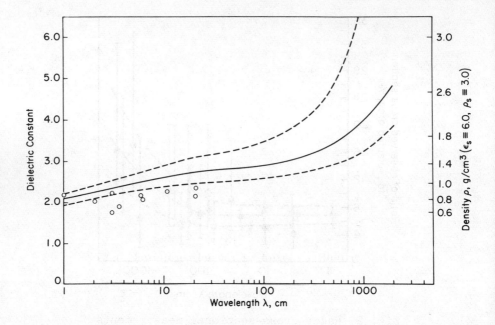

Fig. 3 Estimates of dielectric constant (left-hand scale) and
density (right-hand scale) of the effective reflection
layer. The circles are measurements of ϵ from the
polarization of radio emission.

Table 2 Surveyor radar results

Site	Measured ϵ	Density, g/cm^3
Surveyor I	2.40 ± 0.50	1.0 ± 0.5
Surveyor II	2.07 ± 0.11	0.81 ± 0.1
Surveyor V	2.00 ± 0.16	0.79 ± 0.1
Surveyor VII	3.28 ± 0.40	1.6 ± 0.2

Estimates of ϵ from polarization measurements of radio
emission are also shown in Fig. 3 as circled points. We have
corrected the values of ϵ obtained by the various investigators
for surface roughness by a method developed by Hansen and
Muhleman.[10]   Radar backscatter observations at a given wavelength are well represented by the function (for the normalized
cross section per unit projected surface area)[17]

$$S(\Phi) = \alpha^3/(\sin \Phi + \alpha \cos \Phi)^3 \qquad (35)$$

where α is a mean slope parameter that is strongly wavelength-dependent for the moon and Φ is the angle from the subearth point to the reflecting element. Most of the returned power comes from a small range of Φ, and we can approximate $S(\Phi)$ as

$$S(\Phi) = \cos^{-3}\Phi\, e^{-(3/\alpha)\Phi} \qquad (36)$$

It has been shown by Hagfors[7] that the scattering law is in the form of Eq. (36) in the geometrical optics approach if $\exp(-3\Phi/\alpha)$ is the probability density of tilt angles of flat facets; i.e., most of the radar echo can be explained if local tilt angles are exponentially distributed with an angle $\delta = \alpha/3$ corresponding to e^{-1}. Hansen and Muhleman obtained values of δ as a function of wavelength from the radar data. Monte Carlo calculations were then performed over the moon to compute the depolarization for emission by drawing values of Φ from a table of exponentially distributed random numbers with the appropriate δ; i.e., the local values of θ_0 in Eqs. (2) and (3) were perturbed by random samples of Φ at each integration grid point. The experimental results so corrected are shown in Fig. 3. Estimates of ϵ as a function of λ do not increase as rapidly as do the radar values. This effect is probably due to the polarization effects which occur near grazing incidence (toward the limb) and the $\sec \theta_0$ effect which causes the effective emission layer to be nearer to the surface than the effective normal incidence radar reflection layer at the same wavelength.

Finally, the electrical parameters of an Apollo 11 bulk sample have been measured by Gold et al.[5] at a frequency of 450 MHz. They report values of density, dielectric constant, and absorption length $\ell_{R,\nu}$ in wavelengths. Their results are given in the first three columns of Table 3 (as read from their published graphs).

We have computed the loss tangent with Eq. (27) and the ratio $\tan\Delta/\rho$. It can be seen from the table that the value of $\tan\Delta$ for $\rho = 1$ g/cm^3 measured by Gold et al. is in remarkable agreement with our value given in Eq. (30) computed from the linear theory of the lunar radio emission. Values of $\tan\Delta/\rho$ are nearly constant, as is expected from laboratory measurements.

We shall now briefly discuss the range and variation of the thermal parameters. This subject is treated in Chap. 3a and

Table 3 Apollo 11 results

Density, g/cm^3	ϵ	$\ell_{R,\nu}$, wavelengths	tan Δ	tan Δ/ρ
1.0	1.8	18.0	0.0066	0.0066
1.22	2.0	13.5	0.0083	0.0068
1.55	2.45	10.0	0.0120	0.0077

3b, and we shall limit our discussion to background information for the review of recent model calculations of radio
emission characteristics. It was realized very early that the
thermal conductivity of the upper few millimeters of the lunar
surface under vacuum conditions must be very low. Apparently,
Wesselink[26] was first to point out the importance of the
variation of k with temperature. Krotikov and Troitsky[13]
developed these ideas further. Watson[25] performed important
laboratory experiments with glass beads of various particle
sizes and other materials in vacuum conditions. He demonstrated the variation of specific heats with temperature and
represented the thermal conductivity in the form

$$k = A + BT^3 \tag{37}$$

where the A term arises from contact conduction and the B
term from radiative heat transfer between the grains. The A
term was found to be a function of mineral type, and the B
term was primarily a function of particle size. Recently,
Fountain and West[3] have investigated several materials,
including basalts, under conditions closely simulating the
lunar surface. Some of these results are shown in Fig. 4,
where it can be seen that Watson's form represents the temperature variation very well. Fountain and West also reported
the result of introducing a small amount of CO_2 gas (7 mb) in
simulating the circumstances on Mars. The temperature
variation is then negligible, apparently because of the small
amount of CO_2 in the spaces between the grains. (Undoubtedly
the same thing would occur for other atmospheric gases, e.g.,
air.) Figure 4 also demonstrates the effect of the density of
the basalt; increasing density increases the contact area of
the grains but has little effect on the form of the temperature variation. For the case with ρ = 0.98 g/cm^3 at a temperature of 300°K, the thermal conductivity for the conduction
and radiation parts is 0.595×10^{-5} and 0.465×10^{-5} w/cm °K,
respectively, i.e., essentially the same. Finally, the data

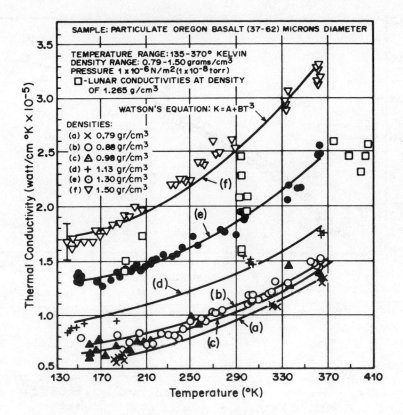

Fig. 4 Thermal conductivity of particulate basalt as a function of temperature and density.

on specific heat variations reviewed by Hagfors[8] are shown in Fig. 5. These data indicate that c_p varies from about 0.2 to 0.8 joules/g-°K over the temperature range appropriate to the moon. It can be seen that these effects are far from negligible!

Models with Variable Parameters

We have seen that models with constant parameters yield a very satisfactory explanation of lunation radio emission observations. However, other observations cannot be adequately explained, such as the details of eclipse cooling which can be observed in the millimeter wave region and, most importantly of all, the increase in the <u>mean</u> brightness temperature over lunations with wavelength. These mean temperatures have been carefully measured by Soviet investigators and are shown in Fig. 6. Obviously, these data suggest a true

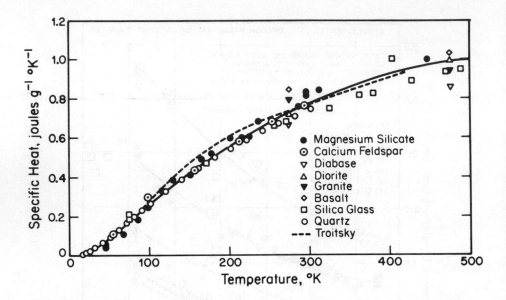

Fig. 5 Temperature variation of specific heats.

increase in the mean temperature with depth into the lunar
surface. Krotikov and Troitsky[13] interpreted this temperature
gradient in terms of radiogenic heating and estimated quantities of radioactive materials much greater than that found on
the earth's crust. On the other hand, Linsky[14] attempted to
explain these data in terms of variable thermal parameters.
As we have seen, the thermal parameters do vary by factors of
2 or 3 over the lunar temperature range, a complex variation
of density with depth certainly exists, and, consequently, all
of the fundamental parameters vary sufficiently to affect the
observed phenomena. Consequently, a number of people have
attempted calculations that more or less include the important
variations. None of these calculations can be regarded as
definitive for various reasons, and, furthermore, various
calculations are difficult to compare, since each worker has
begun with a specific set of assumptions.

The temperature distributions with depth at any specified
point on the lunar surface can be obtained to any desired
accuracy from the numerical integration of the heat equations
with appropriate boundary conditions:

$$c_p(x,T)\,\rho(x)\,\frac{\partial T(x,t)}{\partial t} = \frac{\partial}{\partial x}\left[k(x,T)\,\frac{\partial T(x,t)}{\partial x}\right] \qquad (38)$$

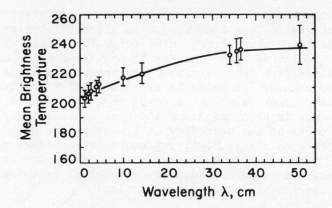

Fig. 6. Increase in mean lunation temperature
with wavelength.

subject to the boundary condition

$$k(x,T) \left[\frac{\partial T(x,t)}{\partial x} \right]_{x=0} = (1 - A_1)T^4(0,t) - (1 - A_B)I(p,t) + Q$$

(39)

where A_1 is the appropriate infrared albedo, A_B the bolometric
albedo (i.e., over the solar spectrum), $I(p,t)$ the insolation
function of the solar radiation at a specified point on the
lunar surface at coordinate p, and Q is the heat flow density
from the moon's interior due both to radiogenic heating and
the primordal cooling of the moon. Clearly, definitive calculations require a knowledge of parameter variations with x,
which are unknown. Furthermore, the temperature curves so
obtained must then be integrated over depth using equations
analogous to Eq. (1), where the dependence on x of the electrical parameters must be known. Definitive calculations
would also require a considerably better understanding of the
effects of surface roughness on both the thermal and the radio
phenomena. Nevertheless, the computations of Linsky[14] and
Troitsky's group do add considerably to our understanding of
lunar phenomena.

Linsky, using reasonable values of the thermal parameter γ,
evaluated at 350°K and a T^3 variation of k, has shown that
differences exist between the <u>mean</u> surface temperature and the
asymptotic temperature at depth between 20 and 50°K. This is

adequate to explain the observed temperature gradient in the
wavelength region from 1 mm to 3.2 cm without interior heating.
The higher temperature differences occur for relatively higher
ratios of radiative heat transfer to contact conduction, and
the variation with temperature of the specific heat is of
little importance. In order to explain the temperature
gradient at longer wavelengths (3-50 cm), Linsky used a heat
flow from the interior of 3.4×10^{-7} cal/cm^2-sec, which is of
the same order as the heat flow in the earth's crust. Several
of Linsky's models adequately represent the observed mean-
temperature data. Expressing the thermal parameter at 350°K
as γ_{350} and the ratio of radiative to contact heat transfer as
R_{350}, he reports, in part, the following:

1) $\gamma_{350} = 885$, $R_{350} = 1$, $\epsilon = 2.50$, heat flux $= 3.4 \times 10^{-7}$
cal/cm^2-sec.

2) $\gamma_{350} = 1030$, $R_{350} = 0.5$, $\epsilon = 1.50$, heat flux $= 3.4 \times 10^{-7}$
cal/cm^2-sec.

The observed temperature gradient and the heat flow figure
cannot extend very deeply into the surface, since the melting
temperatures would be exceeded quickly. It is highly unlikely
that the figure of the moon could be maintained if a significant fraction of the interior were molten.

Linsky's calculated lunation brightness temperatures agree
well with the observations, which are shown in Figs. 7-9. It
can be seen that the observations themselves cannot discrimi-
nate between his various models, which range in γ_{350} from
625 to 1075. Eclipse observations and model calculations are
shown in Fig. 10. In this case, the observations are not very
consistent, but the theory does correctly predict the general
characteristics. Infrared observations are much more sensitive
to γ in the night part of lunations and during eclipses,
although these measurements refer to the upper few millimeters
of the surface.

All of these matters have been reconsidered by the Soviet
investigators (e.g., see the review by Troitsky and
Tikhonova[24]). In particular, they have computed models with
earth-like quantities of radiogenic heating along with temperature variations in the thermal parameters. Mean temperature gradients with wavelength are shown in Fig. 11 for two-
layer models, with an upper layer of thickness d. Troitsky
and Tikhonova state that the temperature increase stops at
wavelengths from 25-30 cm, apparently because the emission is
effectively coming from compacted layers; i.e., the longer
wavelength emission arises from essentially the same depth.

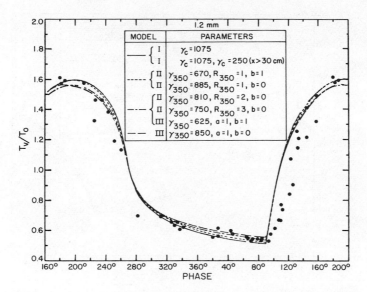

Fig. 7 Computed radio brightness temperatures
at 1.2 mm for the center of the disk
are compared with the data of Low and
Davidson.[15] Figure from Linsky.[14]

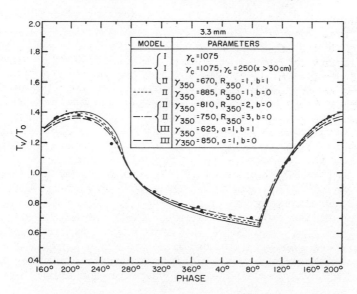

Fig. 8 Computed radio brightness temperatures
at 3.3 mm for the center of the disk
are compared with temperatures inter-
polated from the isotherms of Gary et
al.[4] Figure from Linsky.[14]

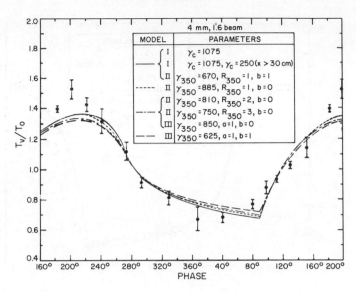

Fig. 9 Computed radio brightness temperatures at
        4 mm for the center of the disk are com-
        pared with the data of Kislyakov and
        Salomonovich.[12] Figure from Linsky.[14]

The results of Fig. 11 are in good agreement with Linsky's
calculations in that they suggest a similar interior heat flux,
which, in turn, is consistent with the earth's flux. Radiogenic heating in the earth's crust arises primarily from the
decay of potassium, uranium, and thorium in the mantle.
According to MacDonald[16], reasonable values for the earth's
crust are K (2.6%), U (2.3 × 10^{-6} g/g), and Th (8 × 10^{-6} g/g),
which he uses to compute a radiogenic heat flux of 0.8 × 10^{-6}
cal/cm^2-sec, whereas measured values on the earth are about
1.2 × 10^{-6}. Very recently, some information became available
concerning the abundances of radiogenic materials in the lunar
Apollo samples. O'Kelley et al.[19] have reported K (0.11%),
U (0.49 × 10^{-6} g/g), and Th (1.92 × 10^{-6} g/g) from Apollo 11
samples. Preliminary analysis of Apollo 12 samples (see Chap.
3b) gives values for, e.g., sample 12070, of K (0.206%),
U (6 × 10^{-6} g/g), and Th (1.5 × 10^{-6} g/g). Although it is
certainly not possible for us to compute a heat flux for the
lunar crust from such meager information, we can state that one
would expect values "similar" to that for the earth.

Clearly, it would be of great importance if the temperature
gradients on the moon could be accurately measured and pre-
cisely interpreted with radio techniques (also see Chap. 2c
for in situ measurement of lunar thermal gradients). Perhaps

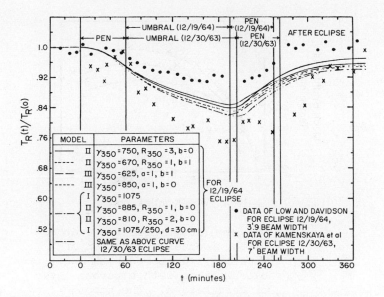

Fig. 10   Computed radio brightness temperatures
          for 1.2 mm at the center of the disk
          for the total lunar eclipse of Dec. 30,
          1963, and Dec. 19, 1964, are compared
          with the data of Low and Davidson[15] and
          Kamenskaya et al. The circumstances of
          these eclipses are sufficiently similar
          that the two sets of data should agree.
          Figure from Linsky.[14]

this alone offers motivation for computing more definitive
thermophysical models of the moon.

Conclusions

Thermophysical models of the moon can be constructed which
explain the radio emission from the moon to the accuracy of
the measurements. The variations during lunations of the
radio brightness temperatures measured in the wavelength range
from 1 mm to about 5 cm can be reproduced with homogeneous
models with a thermal parameter in the range from 800 to 1200
cgs units and electrical parameters $\epsilon \approx 2.0$ and loss tangent
≈ 0.008. Allowing for the temperature variation of the thermal parameters as demonstrated from laboratory measurements
with basalts, in particular, has little effect on the lunation
temperatures and predicts an increase of the mean lunation
brightness temperatures with wavelength, which agrees with
observations at short wavelengths. Apparently, the observed
increase of mean temperature at longer wavelengths (3-70 cm)

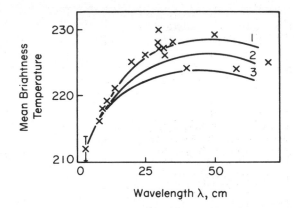

Fig. 11   The mean brightness temperature
          of the moon as a function of
          wavelength compared with model
          calculations with a porous layer
          of thickness d. The smooth
          curves represent: 1) d = 500 cm;
          heat flow = 0.65×10^{-6} cal/
          cm^2-sec, 2) d = 400 cm; heat
          flow = 0.72×10^{-6} cal/cm^2-sec,
          and 3) d = 300 cm; heat flow =
          0.80×10^{-6} cal/cm^2-sec (figure
          from Troitsky and Tikhonova[24]).

can be explained only by hypothesizing a heat flow from the
interior of the moon of from 10^{-7} to 10^{-6} cal/cm^2-sec. Such a
heat flow, if real, is probably caused by radiogenic materials
in the upper kilometers (?) of the lunar crust. Although the
long wavelength measurements are poor, the temperature gradient
appears to reach zero for wavelengths of about 70 cm. There
is no unambiguous way to determine the corresponding depth into
the surface, although models by Soviet workers suggest that
this depth is on the order of 5 m. Certainly, values considerably larger are possible.

   More difficulty is encountered in explaining the millimeter
brightness temperature curves during total lunar eclipses.
This phenomenon is controlled by the upper few centimeters of
the lunar soil (because the time variations are short), and,
consequently, the effects are better studied with infrared
measurements or, perhaps, submillimeter radio measurements.

   The evidence is strong that there is a significant increase
of density with depth. More work needs to be done to

interpret the radar cross-section data in terms of a unique density-with-depth relationship.

Thermophysical models have been developed from earth-based observations alone. These results have proven to be very accurate for all cases where parameters could be compared with Apollo samples, for example. This success allows one to have considerable confidence in the application of the earth-based techniques to the terrestrial planets.

References

[1] Campbell, M. J. and Ulrichs, J., "Electrical Properties of Rocks and Their Significance for Lunar Radar Observations," Journal of Geophysical Research, Vol. 74, No. 25, 1969, p. 5867.

[2] Carslaw, H. S. and Jaeger, J. C., Conduction of Heat in Solids, University Press, Oxford, England, 1959.

[3] Fountain, J. A. and West, E. A., "Thermal Conductivity of Particulate Basalt as a Function of Density in Simulated Lunar and Martian Environments," Journal of Geophysical Research, Vol. 75, No. 20, 1970, p. 4063.

[4] Gary, B., Stacey, J. and Drake, F. D., "Radiometric Mapping of the Moon at 3 Millimeters Wavelength," Astrophysical Journal Supplement, Ser. 12, 1965, p. 239.

[5] Gold, T., Campbell, M. J. and O'Leary, B. T., "Optical and High Frequency Electrical Properties of Lunar Samples," Science, Vol. 167, No. 3918, 1970, p. 707.

[6] Hagfors, T., "Backscattering from an Undulating Surface with Applications to Radar Returns from the Moon," Journal of Geophysical Research, Vol. 66, No. 3, 1964, pp. 777-785.

[7] Hagfors, T., "Relationship of Geometric Optics and Auto-correlation Approaches to the Analysis of Lunar and Planetary Radar," Journal of Geophysical Research, Vol. 71, No. 2, 1966, pp. 379-383.

[8] Hagfors, T., "Remote Probing of the Moon by Infrared and Microwave Emissions and by Radar," Radio Science, Vol. 5, 1970, pp. 189-227.

[9] Hagfors, T., "A Study of Depolarization of Lunar Radar Echoes," Radio Science, Vol. 2, 1967, pp. 445-465.

[10]Hansen, O. and Muhleman, D. O., Journal of Geophysical Research (to be submitted for publication).

[11]Kaydanovsky, N. L., Ihsanova, G. P., Apushkinsky, G. P. and Shivris, O. N., "Observations of Radio Emission of the Moon at 2.3 cm," The Moon, Kopal, Z., ed., Academic Press, New York, 1962, p. 527.

[12]Kislyakov, A. G. and Salomonovich, A. E., "Radio Emission from the Equatorial Region of the Moon in the 4 mm Band," Radiofizika, Vol. 6, 1963, p. 431.

[13]Krotikov, V. D. and Troitsky, V. S., "Detection of the Hot Interior of the Moon," Astronomicheskiy Zhurnal (USSR), Vol. 40, 1963, p. 1076.

[14]Linsky, J. L., "Models of the Lunar Surface Including Temperature Dependent Thermal Properties," Icarus, Vol. 5, 1966, pp. 606-634.

[15]Low, F. J. and Davidson, A. W., "Lunar Observations at a Wavelength of 1 mm," Astrophysical Journal, Vol. 142, 1965, p. 1278.

[16]MacDonald, G. J. F., "Chondrites and Chemical Composition of the Earth," Researches in Geochemistry, Abelson, P. H., ed., Wiley & Sons, New York, 1959, p. 476.

[17]Muhleman, D. O., "Radar Scattering from Venus and the Moon," Astronomical Journal, Vol. 69, 1964, pp. 34-41.

[18]Muhleman, D. O., "Surveyor Project Final Report: Part II. Science Results," TR 32-1265, Jet Propulsion Laboratory, Pasadena, California, 1968.

[19]O'Kelley, G. D., Eldridge, J. S., Schonfeld, E. and Bell, P. R., "Proceedings of the Apollo 11 Lunar Science Conference," Geochimica et Cosmochimica Acta, Supplement 1, Vol. 2, 1970, p. 1407.

[20]Piddington, J. H. and Minnett, H. C., "Microwave Thermal Radiation from the Moon," Australian Journal of Scientific Research, Series A2, 1949, p. 63.

[21]Troitsky, V. S., "Radio Emission of the Eclipsed Moon," Astronomicheskiy Zhurnal (USSR), Vol. 42, 1965, p. 1296.

[22]Troitsky, V. S., "Investigation of the Surface of the Moon and Planets by the Thermal Radiation," Radio Science, Vol. 69D, No. 12, 1966, p. 1585.

[23]Troitsky, V. S., Burov, A. B. and Aloyshina, T. N., "Influence of the Temperature Dependence of Lunar Material Properties on the Spectrum of the Moon's Radio Emission," Icarus, Vol. 8, 1968, p. 423.

[24]Troitsky, V. S. and Tikhonova, T. V., "Thermal Radiation from the Moon and the Physical Properties of its Upper Mantle," Radiofizika, Vol. 13, No. 9, 1970, pp. 1273-1311.

[25]Watson, K., "Thermal Conductivity Measurements of Selected Silicate Powders in Vacuum from 150-350°K," Ph.D. Thesis, California Institute of Technology, 1964.

[26]Wesselink, A. J., "Heat Conductivity and the Nature of the Lunar Surface Material," Bulletin of the Astronomical Institutes of the Netherlands, Vol. 10, 1948, p. 351.

[27]Winter, D. F. and Saari, J. M., "A Particular Thermophysical Model of the Lunar Soil," Astrophysical Journal, Vol. 156, 1969, pp. 1135-1151.

RADAR MAPPING OF LUNAR SURFACE ROUGHNESS

Thomas W. Thompson[*]

Jet Propulsion Laboratory, Pasadena, Calif.

and

Sidney H. Zisk[+]

M.I.T. Haystack Observatory, Westford, Mass.

Nomenclature

b = lunar radius, km T = radar flight time, sec

c = velocity of light, km/sec θ = angle of incidence

t = echo delay time, sec λ = wavelength, cm

Δt = transmitter pulse, sec

Introduction

Both surface and near-surface rocks with sizes in the centimeter and meter range control the radar scattering and thermal emission from the lunar surface. Rocks smaller than about one meter have been photographed only at the Surveyor and Apollo landing sites. For the remainder of lunar surface, only the radar and infrared maps (described in Chapter 1a) have provided information about lunar surface conditions in this size range. Whereas the infrared emission is controlled by bare surface rocks, radar scattering is controlled by both surface rocks and rocks buried near the surface. Thus, the radar and infrared measurements are complementary, and a review of lunar measurements is appropriate to this volume.

This paper presents results from portions of the lunar research conducted at the Jet Propulsion Laboratory, California Institute of Technology, under Contract NAS7-100, and at Haystack Observatory under Contract NAS9-7830. Northeast Radio Observatory Corporation operates Haystack under agreement with Massachusetts Institute of Technology with support of NSF, NASA, AND ARPA.

[*]Member of technical staff.

[+]Staff scientist.

Radar measurements (like thermal measurements) can be divided
into two broad classes: average characteristics and local
departures from the average. In this article the high-
resolution mapping of these departures from the average will be
emphasized. The observed variations in radar scattering could
arise from variations in bulk electrical properties of dielectric constant, permittivity, and conductivity, which are dependent upon surface density. However, the simultaneous observations of radar echoes in different polarizations indicate that
the observed differences are controlled primarily by surface
roughness. Thus, radar maps of the lunar surface depict areas
of roughness with scale sizes near the radar wavelength, presumably indicating surface and near-surface rock fragments.
Surface rocks referred to in Chapter 1.a of this volume, presumably cause the infrared anomalies observed during lunar
eclipses.

The techniques of mapping lunar radar echoes and the interpretation of the results are based solidly on the earlier measurements of average backscattering characteristics of the
moon. The techniques, results, and interpretation of these
measurements of the average lunar characteristics are briefly
reviewed in the next section. The following two sections discuss the techniques and results of radar mapping, and mapping
results at 3.8-cm and 70-cm wavelengths will be discussed in
some detail. The last section reviews a comparison of infrared
and radar mappings and the interpretation of these comparisons
in terms of surface rock populations.

The Average Scattering Behavior of the Moon: A Brief Review

Infrared measurements of the moon rely upon emission of
absorbed solar radiation. In contrast, radar observations are
performed by illuminating the moon with specific forms of electromagnetic energy, and measuring the energy scattered by the
lunar surface. Nearly all radar observations of the moon are
monostatic, where the same antenna is used for transmitting and
receiving, and only energy reflected straight back toward the
observer is measured. However, with the advent of spaceflight,
bistatic radar observations have been performed with the transmitter aboard the spacecraft and the receiver located on Earth.
Here, energy scattered obliquely can be measured; but to date,
only specular forward scattering has been measured because of
low signal strengths. These bistatic measurements have measured the slope of areas within 50 to 100 km of the specular
reflecting point,[1] and measured the moon's dielectric constant
from nulls observed at the Brewster angle.[2] Also, a few monostatic radars have been operated from spacecraft.[3,4] However,
radar mapping has only been done with Earth-based monostatic
radar systems.

The moon's radar-scattering behavior can be measured in two
ways. For example, if a short pulse of energy is transmitted,
then the extended depth of the moon spreads the echo in time.
The earliest echoes are reflected from the area of the moon
closest to the radar, and the latest echoes are reflected
(albeit weakly) from the limb. The time difference between the
earliest and latest echoes is 11,600 μsec, and modern planetary
radars generally have time resolution capabilities of a few
microseconds.

Although these pulse measurements are generally used, equivalent information about the moon's scattering behavior has been
obtained by measuring the spread in echo frequency caused by
the doppler effect. The echoes have a gross doppler shift
imparted by the changing range between the radar and the center
of the moon. In addition, the echoes have a small spectral
spread imparted by the moon's libration. A measure of the
amount of libration is the velocity difference between the east
and west limbs; this velocity difference can be as large as
10 — 15 km/h. This yields doppler-frequency spreads of about
$(700/\lambda)$ Hz, where λ is the radar wavelength in centimeters.
These spectral spreads can be resolved to 0.1–0.01 Hz with
present-day radars.

Average Radar Behavior by Short Pulse Measurements

Consider then, that the moon is illuminated by a radar which
transmits a short burst of energy with time duration Δt.
After a flight time of about 2.5 sec, the echo is observed.
Because of the extended depth of the moon, the echo's duration
is 11,600 μsec. A signal arriving at a time t after the first
echo gives the integrated power from an annulus on the moon's
surface lying at radar ranges between $ct/2$ and $c(t + \Delta t)/2$ from
the subradar point, where c is the velocity of light. The
geometry is shown in Fig. 1; and the subradar point is the
point on the moon closest to the radar.

As a consequence of transmitting short pulses, the annulus
with the echo at time t has the same angle of incidence with
respect to the radar's line-of-sight. Also, the entire reflecting area is independent of the range of the scattering area
(if the entire moon can be illuminated by the radar beam). The
size of this scattering area depends only upon the duration of
the transmitter pulse and is given by $(\pi bc \Delta t)$, where c and Δt
are defined previously, and b is the lunar radius. The amount
of scattering area is several thousand square kilometers, even
for pulses of a few microseconds' duration. Thus, this type of
measurement averages the reflections from a large amount of the
lunar surface.

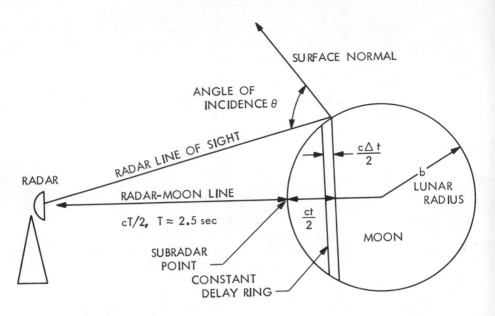

Fig. 1   Scattering area for short-pulse power-versus-delay
observations. For a transmitted pulse of time
duration Δt, the constant delay ring has echoes
at a time $T + t$ after the transmission of the pulse.
All elements of the constant delay ring have the
same orientation with respect to the radar, and
are viewed with an angle of incidence θ. Also,
the constant delay ring has an area of $\pi bc\Delta t$ and
has a width along the surface of $\left[c\Delta t/2 \sin(\theta) \right]$.
The geometry shown here is exaggerated, since the
radar-moon distance is 58 − 60 lunar radii.

As a result of the effects described in the last paragraph,
echoes from short pulses of energy give the backscattered power
per unit area as a function of the angle of incidence and averaged over large portions of the lunar surface. The results of
these measurements, the mean radar-reflective behavior of the
moon, is often called the scattering law of the lunar surface.

The Mean Scattering Measurement: Results

In controlling the form of the energy transmissions, the
radar experimenter can set the polarization of the energy
bursts to make a study of the depolarizing effects of the lunar
surface. A thorough study of these effects requires transmission and reception of all combinations of circular and
linear polarization, and only one experimenter has accomplished
this.[5] However, for the present description of mean scattering
behavior and presentation of mapping results to follow, only
transmission and reception of circular polarization will be
considered. Backscattered signals are often received simultaneously in opposite circular polarization, which are generally
called the <u>polarized</u> and <u>depolarized</u> components of the echo.
The polarized component is the polarization received from a
reflection from a smooth surface; the depolarized component is
the polarization in the sense opposite to the polarized
component.

Although these polarized and depolarized echoes behave quite
differently for areas near subradar point, they behave somewhat similarly for the limb areas. For areas near the subradar point, polarized echo strengths are quite strong and
decrease sharply with increasing angle of incidence. However,
at angles of incidence of about 35°, the angular dependence of
polarized echoes changes to more gradual decay. From 35° to
about 80° angle of incidence, polarized echo strengths (in
backscattered power per unit area) vary as $\cos^{3/2} \theta$ where θ is
angle of incidence. For angles of incidence between 80° and
90°, the dependence approaches $\cos \theta$. In contrast, the depolarized echo strengths vary as $\cos \theta$ from the subradar point
outward. The ratio of polarized to depolarized power is about
1000:1 at the subradar point to about 2:1 at the limbs. These
general comments apply to radar wavelengths from a few centimeters to about 10 m. Although scattering at the longer wavelengths is characterized generally by somewhat strong dependence upon angle of incidence at the subradar point and by
slightly higher ratios of polarized to depolarized powers.

Plots of observed scattering behavior (versus θ) are given in
Figs. 2 and 3. Polarized powers are shown in Fig. 2 for wavelengths of 3.8, 23, and 68 cm. These values are adjusted so
that the integrated power from the entire visible surface gives
a radar cross section of $0.07 \pi b^2 = 6.6 \times 10^5$ km^2, the

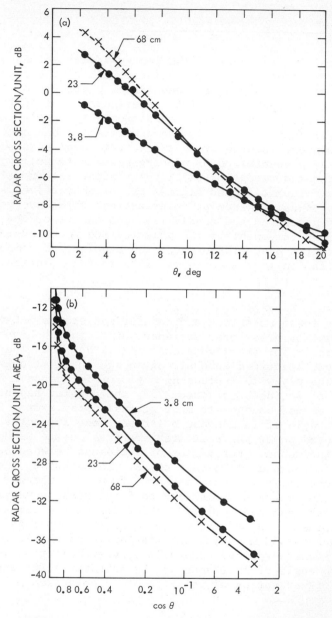

Fig. 2   Average radar scattering of the moon versus angle
of incidence θ. (a) Specular behavior of polarized
radar echoes for wavelengths of 3.8, 23, and 68
cm, (b) Polarized echo power for large angles of
incidence. Abscissa is logarithm of cos θ which
emphasizes behavior at larger angles of incidence
(taken from Figs. 5 and 6 of Ref. 21).

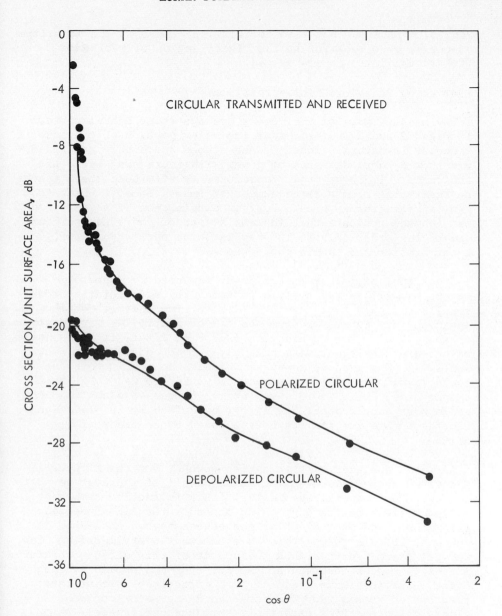

Fig. 3 Average scattering behavior of polarized and
depolarized radar echoes at 23-cm wavelength
(taken from Fig. 7 of Ref. 21).

cross section at those wavelengths. Polarized and depolarized
echo strengths at the single wavelength of 23 cm is shown in

Fig. 3. Since the abscissa for Figs. 2b and 3 is the logarithm of cos θ, echo behavior at the higher angles of incidence (toward the limb) is emphasized.

Scattering Behavior: Interpretation

The measurements of the moon's average radar behavior shown in Figs. 2 and 3 indicate that the major portion of the echo is specular in nature. That is, the echoes are highly polarized and show a sharp decrease of power with increasing angle of incidence. It is natural, therefore, to attribute these echoes to scattering from a large number of smooth facets, which are many wavelengths in size. Further examination of these specular echoes[5] indicate that the rms slopes of the lunar surface are on the order of 5 to 10°, as is borne out by Surveyor, Lunar Orbiter and Apollo photographs.

The amount of backscattered power measured by monostatic radars as well as radiometric (Chapter 1b) and bistatic radar results indicate that "average" dielectric constant for the lunar surface increases from about 2 at centimeter wavelengths to about 4 at meter wavelengths (see Fig. 27 of Ref. 6). These values are consistent with bulk dielectric constants of rock dusts.[7,8] The gradual increase in bulk dielectric constant could result from a large number of buried rocks "seen" by the deeper penetrating radio waves at the longer wavelengths. A radio wave should penetrate into the surface by 10 − 50 wavelengths before its amplitude is reduced by an order of magnitude.

Although specular scattering can account for the effects described above, it cannot account for all of the average scattering. Here we will assume that there is another component called diffuse scattering, which accounts for all of depolarized echoes and most of polarized echoes beyond angles of incidence of 40°. Although this is a somewhat oversimplified view, we will assume further that the source of this diffuse scattering is the rocks observed at the Surveyor and Apollo landing sites. Surface rocks would be more effective as scatterers than buried rocks. Surface rocks have a greater contrast in electrical properties with their surroundings (space), than buried rocks have with their surroundings (the lunar soil). Also, the radar power refracted into the surface is attenuated at a rate of 10 dB for every 10 − 50 wavelengths; so only rocks buried near the surface can contribute to diffuse scattering. Several authors [9,10,11] have shown that there are enough rocks in the Surveyor photographs to account for the observed diffuse scattering.

The size of the rocks is also important. Rocks smaller than
the wavelength would contribute little backscatter, whereas
rocks considerably larger than a wavelength are observed to be
smooth and would scatter specularly. We will assume arbitrar-
ily that only rocks with sizes between 0.25 and 10 wavelengths
will scatter diffusely.

Thus, if surface and near-surface rocks are assigned as the
source of diffuse scattering as we have suggested, and if diffuse scattering is the major source of depolarized echoes, then
the strength of depolarized echoes is directly related to the
population of surface and near-surface rocks in the appropriate
size range.

Delay Doppler Mapping Technique

As was mentioned previously, radar observations of the moon
fall into two classes: the measurement of average properties
described in the previous section, and the mapping of departures from the average which will be described in the next two
sections. The technique of mapping will be presented in this
section, and the results of the mapping will be presented in
the following section.

Simultaneous Frequency and Delay Resolution of Radar Echoes

In the previous section, the discussion of the measurement of
the moon's average scattering properties indicated that lunar
radar echoes are dispersed in both time-delay and frequency.
The dispersion in time-delay, which was caused by the moon's
extent in depth (i.e., along the line-of-sight from the radar)
was measured with the use of short pulses of energy. In practice, many short pulses are transmitted and the measurement of
echoes are carefully timed so that the same reflecting area
always occurs at the same time-delay sample. However, libration introduces doppler shifts, so different elements of this
reflecting area have different frequencies. A mapping of the
strength of the radar echoes can be obtained if this spread of
doppler frequencies can be resolved.

In order to resolve this spread in doppler frequencies, short
pulses with a spectrally pure carrier are transmitted. That is,
the transmitted signal is derived from a primary frequency
standard which is switched to the transmitting antenna for only
short time periods. The spectrum of the echoes is then calculated by a Fourier analysis of consecutive echoes. Modern
radar telescopes are capable of frequency resolutions of 0.1 to
0.01 Hz, which is equivalent to as little as 500 m on the lunar
surface at centimeter wavelengths.

Once the echo has been resolved in time delay and frequency, a given radar resolution element must be related to its corresponding scattering area on the lunar surface. The delay and frequency of any point on the surface can be calculated from the moon's position and velocity in space, and this calculation can be inverted. Unfortunately, there are generally two areas with the same delay and frequency, and and this potential ambiguity must be resolved by one of the techniques to be described below.

When the delay and frequency of a lunar position is computed, it is convenient to use the concepts of an apparent axis of rotation and a libration equator. Briefly, the moon viewed from the radar site appears to be spinning about the apparent axis of rotation; the libration equator lies in the plane that is perpendicular to this apparent axis of rotation at the moon's center of mass.

The geometry of the radar resolution elements with respect to the apparent axis of rotation and libration equator are shown in Fig. 4. Those areas that appear at a constant range are circles centered on the subradar point (assuming the moon to be a sphere), whereas areas that appear at a constant frequency appear as the half-circles which are parallel to apparent axis of rotation. When echoes are simultaneously resolved in time-delay and frequency, the reflecting area is the intersection of these two annuli. Unless these annuli are just tangent, two symmetrically placed reflecting areas appear at the same delay and frequency. This is the ambiguity mentioned previously.

This ambiguity has been resolved by using the small antenna beams, which are characteristic of the larger radar antennas operating at centimeter and meter wavelengths. If these small antenna beams are used, one of the two reflecting areas can be selectively illuminated and observed. At wavelengths of several meters, even the largest antennas do not have small beamwidths. Fortunately, a two-element interferometer can accomplish the same effect, as was demonstrated by Thompson.[12] This type of ambiguity solution was first used for Venus radar observations.[13]

In addition to the delay-doppler technique just described, radar maps of the moon have been made with synthetic aperture techniques, which was invented by J. H. Thomson and applied by several observers.[14,15] These maps use only spectral resolution and observations are made during special days when the angle between moon's actual and apparent axis of rotations rotates by 180°. At the present time, the resolutions obtained

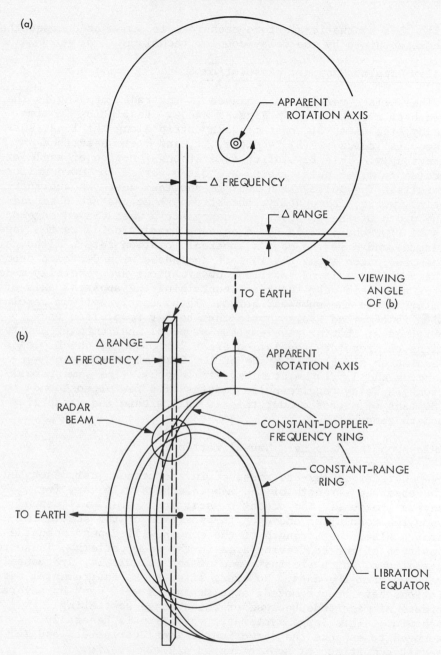

Fig. 4   Geometry of lunar reflecting area for delay-doppler
         mapping: (a) moon as viewed from a point on the
         apparent rotation axis, (b) moon as viewed from the
         plane of the libration equator. This is a modified
         version of Fig. 1 of Ref. 9.

with this synthetic-aperture technique is somewhat coarser than those obtained by the delay-doppler technique.

Delay-Doppler Mapping: Resolution

The surface resolution achieved in the radar mapping is the width of strips shown in Figs. 1 and 4. Resolution in time delay, the width of constant delay strip along the lunar surface, is $[c\Delta t/2 \sin(\theta)]$ where c, Δt, and θ are described previously. This resolution is a strong function of angle of incidence θ; so pulse widths are often varied for more uniform results. The pulse width must be shortest near the subradar point, where, fortunately, the strongest echoes are observed. The width of the constant frequency strip when viewed edge-on is $\Delta f \, \lambda/\Omega$, where Δf is the frequency resolution, λ is the wavelength, and Ω is the moon's apparent rotation rate. Although the resolution along the surface decreases with distance from the apparent axis of rotation, observations are generally made when areas are near the plane containing the apparent axis of rotation and the subradar point. Thus, nearly optimum resolution is obtained. Also, note that better resolution is obtained at shorter wavelengths. Typical instrumental limitations on frequency resolution are $0.1 - 0.01$ Hz, which yields resolution along the surface of $1 - 2$ km at centimeter wavelengths and $5 - 10$ km at meter wavelengths. The same resolutions in delay require pulse lengths of a few microseconds to a few tens of microseconds; this is well within the capability of modern radars.

Delay-Doppler Mapping: Normalization

When radar echoes are resolved in the manner just described, the observed strength of the mapped echoes will vary for several reasons. The amount of scattering area will depend upon the resolution and upon the position of the scattering area. Also, the strength of the echoes will depend upon the position of the scattering area in the beam pattern. These two variations, which are instrumental and calculable, are usually removed from the data. However, after these instrumental variations have been removed, echo strengths will vary by several orders of magnitude because of the average scattering behavior. This large systematic variation is generally removed to enhance the detection of local variations and to permit comparison of scattering on a global basis.

Thus, after these "expected" variations have been removed, the radar maps show only departures of echo strengths from the average.

Radar Mapping Results

The mappings using the delay-doppler techniques previously
described have required sufficient sophistication that only
Earth-based radars have been used. Thus, observations have
been restricted to the wavelengths from a few centimeters to
about 10 m. Shorter wavelengths do not penetrate the water
vapor in the atmosphere, whereas longer wavelengths do not
penetrate the ionosphere. However, radar mappings have covered
most of the observable range. The most extensive and highly
resolved mappings are at 3.8- and 70-cm wavelengths, whereas
coarser mappings have been obtained at 23-cm, 73-cm, and 7.5-m
wavelengths. These mappings are shown in Figs. 5 — 11; radar
operating parameters are given in Table 1.

The division between the coarse and highly resolved mapping
is taken as 10 km. Only the highly resolved maps at 3.8 and
70 cm will be discussed in detail, and there are enough differences between these two maps that each will be described
separately below.

Before describing the scattering differences seen in these
maps, the interpretation of radar maps will be briefly reviewed.
One obvious cause of scattering differences is the occurrence
of local tilts. A tilted area, such as crater rim or mountain
side, is viewed at smaller angle of incidence than its environs
and the mean scattering behavior alone will enhance the echo
strength. These types of changes are very noticeable in the
highlight-shadow appearance of the 3.8-cm radar maps (Fig. 5).
Generally, the scattering differences from tilts occur in
obvious places, and they are extremely useful in locating the
position of a mapped area. The effect of tilts is least pronounced for depolarized echoes from non-limb areas because of
the weak dependence of radar scattering low on angle of
incidence.

Other scattering differences are associated with the
wavelength-sized roughness of the lunar surface, which we
equated to surface and near-surface rocks in our discussion of
the diffuse component of radar scattering. The surface rocks
have a marked effect on the infrared behavior of lunar surface.

Observed Differences at 3.8-cm Wavelength

General. Figure 6 is a map of the depolarized 3.8-cm radar
backscatter from the earthside lunar hemisphere. Several general observations may be made from a first look at the figure.
First, although there are a few large areas of low (dark) radar

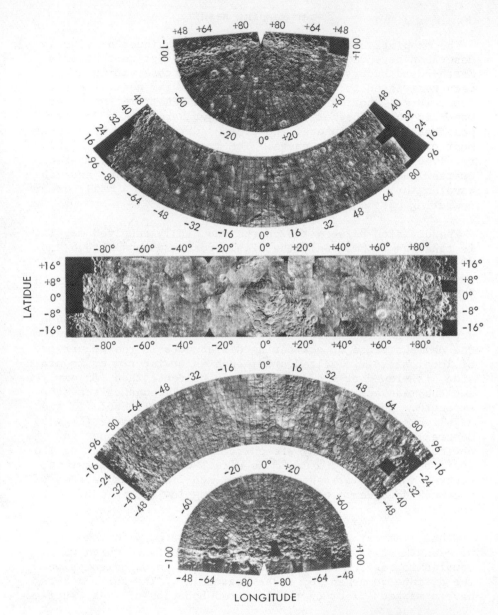

Fig. 5   Mosaic maps of polarized radar echoes at
          3.8-cm wavelength. Polar and mid-latitude
          regions are shown in the Lambert Conformal
          projection; equator regions are shown in
          Mercator projection (taken from Figs. 17
          of Ref. 9).

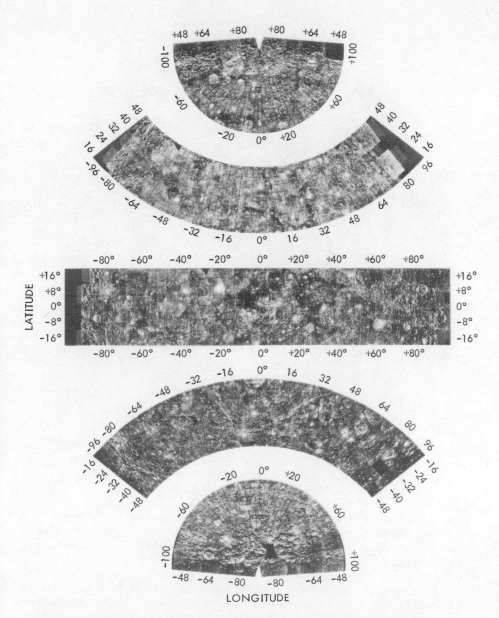

Fig. 6   Mosaic maps of depolarized radar echoes at 3.8-cm
         wavelength.   Polar and mid-latitude regions are
         shown in Lambert Conformal projection; equatorial
         regions are shown in Mercator projection (taken
         from Fig. 17 of Ref. 9).

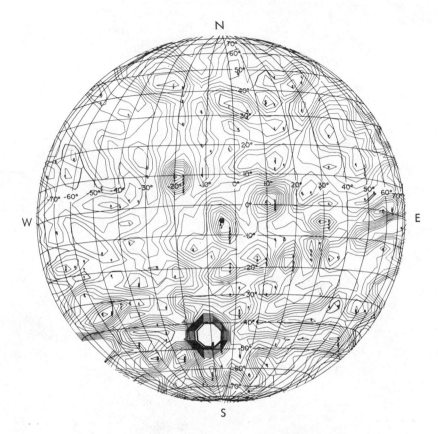

Fig. 7 Map of depolarized radar echoes at 23-cm wavelength, obtained by synthetic aperture techniques. Note enhancements from Tycho and
Copernicus (taken from Fig. 8 of Ref. 14).

backscatter, the most common and most striking features are the
bright areas. Among the bright areas, there appear to be a
number of examples which can be divided into seven classes,
which will be described below.

Also, within about 60° of the mean center of the disk there
is relatively little of the shadowing effect that results from
local tilting of the surface (i.e., crater walls and mountain-
sides). This, as mentioned earlier, is generally true for the
depolarized maps, and is one of the reasons for the usefulness
of the depolarized data. The features in the central region
appear, therefore, to characterize differences in the surface
roughness, rather than in the local topography.

Fig. 8 Mosaic map of polarized radar echoes at
        70-cm wavelength. Earthside hemisphere
        is shown at mean libration, and indi-
        vidual maps are outlined.

Beyond 60° from the center, shadowing effects are again in
evidence, because of **increased dependence of radar scattering**
law with angle of incidence. As a result, it is more difficult
to separate the brightening due to roughness from that due to
tilts. Thus, the number of small bright anomalies found in the
outer regions will be artificially low.

Radar-Bright areas. As mentioned previously, there are at
least seven classes of anomalously bright areas. No attempt
will be made to present comprehensive lists, since the number
of anomalies appears to be in the thousands, and since about

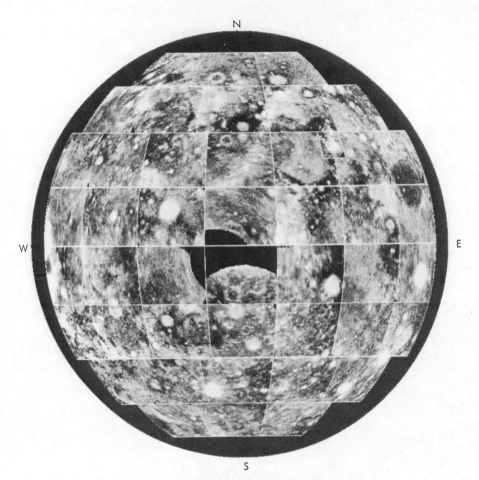

Fig. 9 Mosaic map of depolarized radar echoes at
70-cm wavelength. Earthside hemisphere
is shown at mean libration, and individual
maps are outlined.

30% of them have not been identified with previously named
features. For the first two classes (large bright craters and
small bright craters with bright ejecta blankets), lists of
representative features are given in Table 2. Examples of
named features in the outer categories are given below, and
other examples are given elsewhere.[9,16]

Class 1a: large bright craters (30- to 100-km dia). All of
these have bright walls and/or floors, and show strong scattering from an ejecta blanket. There is gradation in

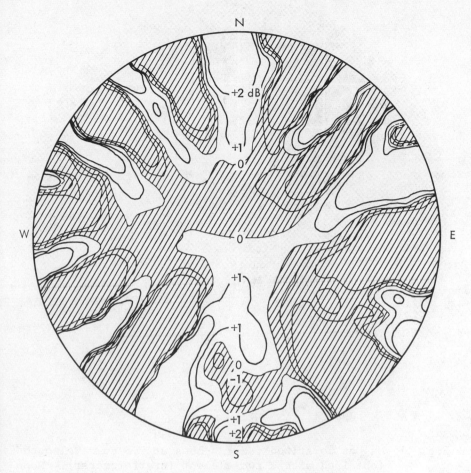

Fig. 10 Map of polarized radar echoes at 73-cm
          wavelength, obtained by synthetic
          aperture technique. Contours show
          departures from average scattering in
          decibels (taken from Fig. 7 of Ref. 15).

brightness from craters like Tycho and Copernicus, which are
essentially immersed in a relatively featureless bright field
of ejecta, down to Zuchius, which has a relatively subdued
ejecta blanket. The dividing line between this and the next
class is somewhat arbitrary.

Class 1b: bright craters with no ejecta (30- to 100-km dia).
Here again there is a gradation from craters like Theophilus,
which might have been included with class 1a (since there is a

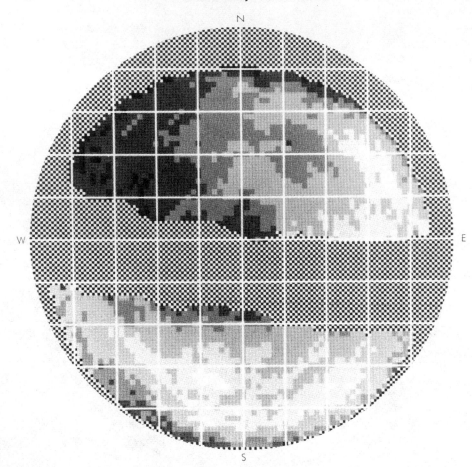

Fig. 11   Map of polarized radar echoes at 7.5-cm wavelength
           (obtained with a two-element interferometer). Moon
           is shown at mean libration. Enhancement from Tycho
           is evident, and depressed echo strengths are ob-
           served for Mare Serenitatis and Oceanus Procellarum.
           Cross-hatched regions could not be resolved, since
           they were near the limb where weak echoes occurred,
           or were near the apparent equator where resolution
           of the interferometer failed.

weak bright area to the north) down to craters like Plato,
which show little more than a bright rim. Craters in this
class, however, have features on their floors (e.g., central
peaks, more recent craters, or "collapse" faults) that are dis-
tinctively brighter than any over-all ejecta field, as opposed

Table 1. Radar mappings of the moon

Figure number	Radar wavelength, cm	Surface resolution, km	Method	Reference
5 and 6	3.8	1 — 2	Delay-doppler/ small antenna beam	9
7	23	50 — 200	Synthetic aperture	14
8 and 9	70	5 — 10	Delay-doppler/ small antenna beam	22
10	73	100 — 300	Synthetic aperture	15
11	750	50 — 200	Delay-doppler/ interferometer	12

to class 1a where the ejecta (presumably rocky rubble) domi-
nates the radar backscatter. Figure 12 is a radar image of
Theophilus, together with the older crater Cyrillus. Note
that Theophilus has a bright central peak, some bright structure on its floor, and a bright rim. Table 2 contains a list
of other class 1b craters.

Class 2. Medium-sized bright craters (10- 30-km dia). There
are a number of smaller bright craters, with ejecta blankets
extending at least a crater diameter beyond their walls. Many
have extremely bright walls or floors. In others, the overall
blanketing is so bright that only the walls show up slightly
brighter. A few (e.g., Proclus, Messier A) have strong rays
that extend a few crater diameters beyond the symmetrical
ejecta blanket. The extended ray structure which appears
optically is almost always absent in the radar maps. This
suggests that most of the rays are of material too finely
divided to generate a radar return even at 3.8-cm wavelength.

Class 3: small bright craters (diameters less than 10 km).
There are a number of smaller craters (10-km dia) that show
radar ejecta blankets two to three crater diameters beyond
their rims (an area 5 to 7 times the diameter of the crater).

(a) POLARIZED

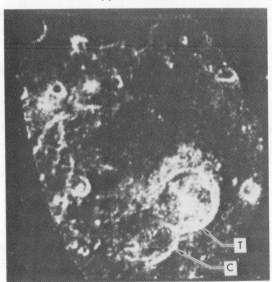

(b) DEPOLARIZED

Fig. 12.   Map of polarized and depolarized radar echoes from
            the craters Theophilus (T) and Cyrillus (C) and
            their environs. Crater Theophilus has the high
            depolarized echoes described in the test.
            Theophilus has a diameter of 100 km and is located
            at 26.3°E and 11.4°S.

Table 2   Examples of large bright craters
at 3.8-cm wavelength

Class 1a Craters with ejecta borders ≤ 1 crater
diameter in width

Name	Location	Remarks
Copernicus	N 10° W 20°	
Tycho	S 43° W 10°	Two or three bright rays
Langrenus	S 8° E 61°	
Stevinus	S 33° E 55°	
Kepler	N 8° W 38°	Possible ray(s)
Aristarchus	N 23° W 46°	
Geminus	N 33° E 56°	
Zuchius	S 63o W 60°	

Class 1b Craters. Large, bright craters with
no obvious ejecta[a]

Plato	N 31° E 30°	w, bf
Posidonius	N 31° E 30°	w, bf, ff
Archimedes	N 30° W 3°	w, bf
Autolycus	N 31° E 3°	w
Theophilus	S 11° E 26°	w, bf
Marius	N 12° W 51°	w
Petavius	S 26° E 61°	bf, ff
Moretus	S 71° W 4°	bf, ff
Gassendi	S 18° W 39°	bf, ff

[a]bf = bright floor.

 w = bright walls and rim.

ff = fractured floor.

In many cases, only the walls of the crater are visible on the
radar map, and in a number of cases only the ejecta blanket
appears. The central crater is visible in the polarized radar
map but is completely overwhelmed by the ejecta in the depolarized map. It is interesting to note that no larger (greater
than 50 km) crater shows this intense radar return from its
ejecta blanket. This fact may result from the greater ages of
all large craters, or may be an effect inherent in the formation of large and small craters.

Class 4: bright diffuse areas with no central crater. There
have also been found a few diffuse, circular enhancements similar in appearance to the "ejecta blankets" discussed previously, but with no central crater visible even on Lunar Orbiter
photography (e.g., Fig. 13). The most likely origin for such a
feature is a clump of very loosely cohesive material ejected at
low velocity from an event some distance away. Radar features
in this class often lie along rays, which is a further argument
in favor of their origin as clumps of ejected material.

Class 5: bright rays. In addition to the crater, and crater-
like, classes just listed, there are several other classes of
radar-bright features. There are two or possibly three major
rays that appear on the 3.8-cm radar map. These rays are associated with the crater Tycho and are located as follows: i) NE
from the crater; ii) NNW from the crater; and iii) a possible
ray cutting across Bessel in Mare Serenitatis. None of these
are visible closer than about 500 km to the crater; they begin
at such a distance, grow somewhat brighter at increasing distances, and then fade away. The Serenitatis ray is not visible
beyond the southern edge of the mare.

The rays visible to the radar presumably contain a larger
fraction of decimeter-sized rocks than their less visible
counterparts. Over a highland terrain such as the area southwest of Mare Serenitatis, either the radar returns are masked
by the general radar brightness of the region or else there is
a physical merging of the ray material with the surface material with the surface material (e.g., mixture into the regolith)
which renders it less visible.

Class 6: mountain peaks. A number of mountain peaks appear
to be visible, notably in the Appenine and Carpathian mountains. There may be some confusion effect from the extreme
topography, but several of the peaks do appear to be somewhat
offset in position from the tilt-induced brightening in the
polarized maps. Hence, the brightness in the depolarized maps
probably results from exposed or near-surface rocks.

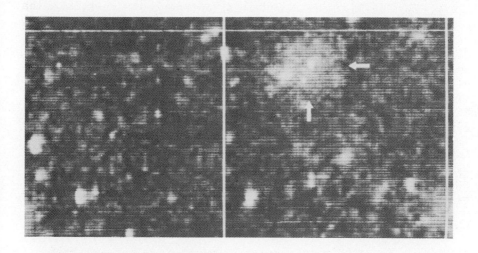

Fig. 13   An area in Oceanus Procellarum southwest of crater
          Kepler. The bright area in the upper-right-hand
          portion of the 3.8-cm radar map has a bright
          nucleus which is located midway between the two
          rightmost 3-km-diam craters on the framelet with
          Kepler E (framelet width is about 10 km). Also,
          see Fig. 33 of Chapter 1a, since this area is also
          an infrared anomaly. (a) 3.8-cm radar map, (b)
          Lunar Orbiter photograph No. IV-144.

Class 7: rilles. A last class of radar-bright anomaly is
the rille. Vallis Schröteri is without peer, in that it is
extremely bright and well-defined and is quite visible on a
darker field. Hyginus rille is almost as bright, but is on a
more confused background. Some rilles (for example, Hadley)
are very weak or invisible, suggesting that they are covered
with a fine-grained mantling layer.

Radar-Dark Areas at 3.8-cm Wavelength. The few areas that
appear anomalously dark should be mentioned also.

Class D-1: mare floors. Mare surfaces are generally darker
than highlands. Some, however, are unusually dark even among
maria. An example is Sinus Aestuum on the eastern edge of
Procellarum. One possible explanation is that they are recent
lava flows and are covered by fine-grained material with few
wavelength-size scatters.

Class D-2: dark halo craters. Definite dark flows appear in
several places, notably at the dark-haloed craters, e.g., on
the floor of Alphonsus. Here again there is an unexpected correlation between the optical and 3.8-cm albedo on a small scale.
Other dark flows appear near fissures at the borders of maria,
for example, at the western edge of Mare Humorum, and even more
extensively at the Sulpicius Gallus formation along the southern
rim of Serenitatis.

Class D-3: dark highland areas. Finally, there are a few
areas that appear on optical photographs to be bright highlands
but are unusually dark on the radar map. An excellent example
is the region to the North of Plato, where the radar-dark
region is punctuated by a few bright craters that are all but
lost on optical photographs.

Observed Scattering Differences at 70-cm Wavelength. Most of
the scattering differences in the 70-cm radar maps are associated with craters. Thus, the scattering characteristics of
craters will be discussed before describing other lunar features having backscattering differences.

Lunar craters having radar enhancements at 70-cm wavelength.
As expected from earlier observations,[17] all of the large rayed
craters had scattering anomalies. The depolarized echoes from
these craters were 10 — 15 times that of the crater's environs.
For polarized echoes, however, the amount of enhancement varied
with the aspect angle of the craters. Craters, such as
Copernicus and Theophilus, which lie near the subradar point,
were observed at low angles of incidence, and had small
enhancements of polarized echoes. Other rayed craters, such as
Tycho, Aristarchus, and Langrenus, lie closer to the limbs and

were observed at higher angles of incidence, and had stronger
enhancements of polarized echoes. However, as a general rule,
these rayed craters backscattered twice as much polarized power
as depolarized power. The area having definite enhancements
often extends beyond these rayed craters by a few rim widths,
but these enhanced areas are much smaller than the rays. In
fact, no rays had enhancements.

Some young, but nonrayed, craters exhibited scattering
enhancements as strong as those of the rayed craters. The
strength of enhancements for other nonrayed craters decreased
to being just detectable. The typical crater having enhanced
radar echoes is young in appearance, has a sharp rim, and is
surrounded often by a hummocky ejecta pattern. Although the
majority of the craters having enhancements were the young
(rayed and nonrayed) craters, a few older craters had weak
radar enhancements.

Several classes of craters had the scattering characteristics
listed in Table 3. For example, 16 craters had radar enhancements and an obvious pattern of large fractures on their
floors. In addition, 8 craters had rougher rims than floors.
Some of these rough-rim craters have floors that are flooded by
mare material. Also 10 craters had enhancements that were
surrounded by a halo of lower returns. This "dark" halo
extended beyond the area of high returns by 1 or 2 crater diameters. Evidently, the ejecta thrown to this area was fine-
grained (smaller than a few centimeters in size) and formed a
smooth blanket over the pre-existing surface.

Other lunar features having radar enhancement at 70-cm wavelength. Besides the craters, other lunar features had distinct
scattering differences. For example, only 5 rilles had enhancements. The strongest of the rille enhancements is associated
with Valli Schröteri, which is located in the Aristarchus
Plateau. Four rilles having weak enhancements were Rima Sharp
I, Rima Hadley, Rima Hyginus, and Rima Aristoteles I. The
latter is a string of craters northwest of the crater Aristoteles. Although other rilles could have enhancements, they are
generally too small to be resolved.

Another lunar feature which had enhanced echoes at 70-cm
wavelength were the ridges and isolated mountain peaks that protrude from the mare surface. One of these features, the Flam-
stead Ring, is probably the remnant of a young crater that was
almost buried by mare material. Other ridges and peaks are
generally attributed to the inner rings of the Imbrium Basin.
The ridges and mountain peaks that exhibited radar enhancements
also are listed in Table 3.

Table 3   Radar features identified at 70-cm wavelength

Craters with enhanced radar echoes and a floor with a fracture pattern	Crater surrounded by a halo of lower radar echoes	Craters with a definite indication of enhanced radar scattering from their rims	Ridges and isolated mountain peaks with enhanced radar echoes	Areas with low radar echoes and appearance in Lunar Orbiter photograph of recent mantling
Maury A	Plato	Plato	Montes Teneriffe ϵ	In Mare Frigoris Northeast of Montes Alpes
Atlas	Plato J	Aristoteles	Montes Teneriffe[a] (α, γ, δ, μ, ι)	On Northwest Shore of Lacus Somniorum
Franklin	Aristoteles	Eudoxus	Pico	West of Mare Humorum
Briggs	Aristillus	Archimedes	Pico β	Western shore of Mare Humorum
Hevelius	Aristarchus	Campanus	Montes Spitzbergensis (μ)	Between Vieta T and La Croix K
Encke	Galilaei A	Pitatus	Montes Spitzbergensis[a] (α, β, γ, θ, κ, E)	Floor of Drebbel E
Taruntius	Galilaei	Gemma Frisius		Flow that meanders Northeast from Drebbel E
Hansteen	Reiner	Maurolycus		
Byrgius D	Theophilus			
Mersenius	Bullialdus			

[a] Adjacent peaks not resolved.

Table 3   Radar features identified at 70-cm wavelength (continued)

Craters with enhanced radar echoes and a floor with a fracture pattern	Crater surrounded by a halo of lower radar echoes	Craters with a definite indication of enhanced radar scattering from their rims	Ridges and isolated mountain peaks with enhanced radar echoes	Areas with low radar echoes and appearance in Lunar Orbiter photograph of recent mantling
Gassendi			Piton	Northern floor of Schickard
Vitello			Unnamed peak in Montes Harbinger	Southeast floor of Schickard
Thebit				
Arzachel			La Hire	Flooded plain southwest of Schiller
Bohnenberger			Flamsteed Ring	
Petavius				

The radar features just mentioned had stronger echoes than their surroundings. However, there are 10 areas of irregular shape that had lower returns than their surroundings. The appearance of these areas in the high-resolution photographs of the Lunar Orbiter spacecraft suggests young mantelling of mare material. That is, they had a low (optical) albedo and a low density of craters. It is most probable that these areas are among the youngest extrusions of mare material. These areas are also listed in Table 3.

Besides these features, mare-highland contacts were generally characterized by higher echo strength from the highlands. There are, however, some exceptions to this generality. For example, the highlands surrounding the crater Plato, and the Jura mountains northeast of Sinus Iridum had the same echo power as the adjacent mare areas.

Besides the mare-highland differences, interesting differences occurred within a mare. For example, a partial ring of low reflecting area occurs on the southern shores of Mare Serenitatis. Also, an area of significant contrast occurs in the northern portion of Mare Imbrium. Schaber, Eggleton, and Thompson[18] showed that the lowest radar echoes in Mare Imbrium are associated with the youngest mare units.

Discussion

The scattering differences in the depolarized 3.8- and 70-cm maps result from surface or near-surface roughness with scales from about 0.25 wavelength to about 10 wavelengths. This roughness could be a rough surface-space interface, rocks lying upon the surface, or rocks buried up to about 50 wavelengths. The ambiguity in type of roughness can be resolved by the infrared observations. Bare surface rocks would have a higher thermal inertia than the dusty surface and appear with a higher temperature during an eclipse. An excess of these surface rocks (with sizes greater than 10 cm) would appear as an infrared anomaly, the "hot spots" discussed in Chapter 1.a of this book. Thus, the infrared maps and radar maps contain complementary information about lunar surface structure with centimeter-to-meter sizes.

A classification scheme for infrared and radar anomalies has been derived by the authors and several colleagues.[19] By definition, an infrared anomaly exhibits increased eclipse temperatures; a radar anomaly exhibits increased depolarized backscatter. An area was simply classified as to whether it was anomalous or not. The eight possible combinations of infrared, 3.8- and 70-cm radar behaviors are those listed in Table 4 and are briefly described as follows:

Table 4. Surface conditions inferred from infrared and radar observations (from Ref. 19)

Anomaly type	Infrared hot spot (increased eclipse temperature)	3.8-cm Radar enhancement	70-cm Radar enhancement	Inferred surface	Relationship to simplified geologic aging model	Occurrence
0	No	No	No	Normal distribution of rocks, undisturbed aged surface	Asymptotic surface, provides basis for definition of anomaly, most widespread	Most common
I	Yes	Yes	Yes	Excess number of centimeter and meter-sized rocks	Predicted for recent cratering event	Common
II	Yes	Yes	No	Excess number of centimeter-sized rocks only: a very young feature	Not predicted	Common
III	No	No	Yes	Older crater covered with a few meters of regolith	Predicted for cratering event of moderate size	Common
IV	Yes	No	Yes	Excess number of meter-sized surface rocks	Not predicted	Rare
V	Yes	No	No	Excess number of smooth surface rocks much larger than meter-size or smooth bare rock surface	Not predicted	Rare
VI	No	Yes	Yes	Top of regolith rough on centimeter and meter scale but no excess of surface rocks	Not predicted	Rare
VII	No	Yes	No	(1) Top of regolith rough on centimeter scale but no excess of surface rocks, (2) alternatively, an excess number of surface rocks in 1-10 cm size range	Not predicted	Rare

The Average Surface: Type 0 Anomaly. The one combination of
behaviors which is not anomalous at any wavelengths represents
the average surface, which covers about 90 — 95% of the Earthside hemisphere. Surface conditions are typified by the Apollo
landing sites (through Apollo 14 at writing) and all Surveyor
landing sites except Surveyor VII. At the latter, the concentrations of bare surface rocks typify the Type I anomaly next
described.

The Most Common Anomaly: Type I Anomaly. The most common
anomaly is classified as an anomaly at all three wavelengths.
This is not surprising, since the infrared anomaly[20] infers a
large number of surface rocks, and these surface rocks would
create anomalies at centimeter and meter wavelengths. Large
numbers of surface rocks are found at young craters where the
ejecta has not been worn away by meteorites.

An Unusual Concentration of Centimeter-Sized Rubble: Type II
Anomaly. The Type II anomaly has average scattering characteristics in the 70-cm radar maps, but anomalous behavior at
3.8-cm radar wavelengths and in the infrared during the lunar
eclipse. The infrared behavior implies an enhanced concentration of surface rocks, whereas the difference in radar behavior
implies that these surface rocks are primarily of centimeter
size. At first thought, this is an unusual combination of
behaviors, since the radar mappings sample areas which are
several tens of kilometers in length. Some event occurred and
generated only centimeter-sized rubble over this large an area.
These anomalies are common and are classified as Class 3 and
Class 4 features in our previous discussion of 3.8-cm mapping
results.

Older Craters: Type III Anomalies. The Type III anomaly has
an enhanced radar echo at 70-cm wavelength, but average
behavior in the 3.8-cm radar wavelengths and average thermal
behavior during an eclipse. This was expected for an older
crater which has been subjected to an intermediate period of
meteoritic bombardment. The area in the uppermost few meters
of the surface has been "gardened" until an average distribution of surface rocks occurs. However, the rubble below these
few meters of "average soil" still contains enough meter-sized
rocks to reflect the 70-cm radar waves which can penetrate
about 10 m into the surface. One crater with these characteristics is Gassendi, on the northern shore of Mare Humorum. If
Mare Humorum has an age of 3.5 billion years, the age of the
mare at the Apollo 11 and 12 landing sites, then the "garden-
ing" in this time period has only penetrated 3 — 5 m of the
surface.

The characterization of anomalies considered only a binary (yes-no) decision on the presence of an anomaly, and did not include another radar characteristic of older craters. Generally, older craters tend to have rough rims. Although the floors of these older craters may have lost most of their radar enhancement, this radar behavior indicates that the floors of craters are smoothed before the rims.

The Rare Anomalies: Anomaly Types IV – VII. The anomalies of Types 0 through IV are common features, whereas the remaining anomalies are rare. The rare anomalies, Types IV through VII, are either infrared anomalies or 3.8-cm radar enhancements, but not both. The presence of the infrared anomaly indicates an enhanced population of surface rocks which must be at least in the centimeter-size range. Thus, it appears that these smaller rocks are not selectively destroyed by the lunar environment for these anomalous areas.

Summary. The comparison of anomalous behavior at infrared and radar wavelengths has permitted an estimation of surface structure with centimeter-to-meter sizes.

Although a simple classification was used, it was demonstrated that the combinaton of the infrared and radar data contained more information than any of the sets by itself. For example, the infrared data indicates that bare surface rocks are the source of the roughness which generates enhanced radar backscatter. The different radar maps indicates the size of surface rocks as was demonstrated for the strewn fields of centimeter-sized rubble, the Type II anomaly. As craters age, they have a rim-bright appearance in the radar maps, and eventually lose their anomalous behavior infrared and 3.8-cm radar wavelengths. The vast majority of the lunar surface has aged to the point where it exhibits no anomalous behavior.

References

[1]Tyler, G. L., and Simpson, R. A., "Bistatic Radar Measurements of Topographic Variations in Lunar Surface Slopes with Explorer 35," Radio Science, Vol. 5, 1970, pp. 263-271.

[2]Tyler, G. L.,"Brewster Angle of the Lunar Crust," Nature, Vol. 219, 1968, pp. 1243-1244.

[3]Brown, W. E.,"Lunar Surface Surveyor Radar Response," J. Geophys. Res., Vol. 72, 1967, pp. 791-799.

4Muhleman, D. O., et al., "Lunar Surface Electromagnetic Properties," Chap. VII in Surveyor Project Final Report, Part II Science Results, Technical Report 32-1265, Jet Propulsion Laboratory, Pasadena, Calif, 1968.

5Hagfors, T., "A Study of the Depolarization of Lunar Radar Echoes," Radio Science, Vol. 2 (New Series), 1967, pp. 445-465.

6Hagfors, T., "Remote Probing of the Moon by Infrared and Microwave Emissions and by Radar," Radio Science, Vol. 5, 1970, pp. 189-227.

7Campbell, M. J., and Ulrichs, J., "The Electrical Properties of Rocks and Their Significance for Lunar Radar Observations," J. Geophys. Res., Vol. 68, 1969, pp. 423-447.

8Gold, T., Campbell, M. J., and O'Leary, B. T., "Optical and High-Frequency Electrical Properties of the Lunar Sample," Science, Vol. 167, 1970, pp. 707-709.

9Lincoln Laboratory, "Radar Mapping of the Moon, Final Report," Lincoln Laboratory, Lexington, Mass., Feb. 28, 1970.

10Burns, A. A., "Diffuse Component of Lunar Radar Echoes," J. Geophys. Res., Vol. 74, 1969, pp. 6533-6566.

11Thompson, T. W., et al., "Radar Maps of the Moon at 70-cm Wavelength and Their Interpretation," Radio Science, Vol, 5, 1970, pp. 253-262.

12Thompson, T. W., "Map of Lunar Radar Reflectivity at 7.5 m Wavelength," Icarus, Vol, 13, 1971, pp. 363-370.

13Rogers, A. E. E., and Ingalls, R. P., "Radar Mapping of Venus with Interferometric Resolution of the Range-Doppler Ambiguity," Radio Science, Vol. 5, 1970, pp. 423-433.

14Hagfors, T., Nanni, B., and Stone, K., "Aperture Synthesis in Radar Astronomy and Some Applications to Lunar and Planetary Studies," Radio Science, Vol. 3 (New Series), 1968, pp. 491-507

[15] Thomson, J. H., and Ponsonby, J. E. B., "Two Dimensional Aperture Synthesis in Lunar Radar Astronomy," Proc. Roy. Soc., A, Vol. 303, 1968, pp. 477-491.

[16] Lincoln Laboratory, "Radar Studies of the Moon," Final Report, Vol. 2, Lincoln Laboratory, Lexington, Mass., 1967.

[17] Thompson, T. W., and Dyce, R. B., "Mapping of Lunar Radar Reflectivity at 70 Centimeter," J. Geophys, Res., Vol, 71, pp. 4843-4853, 1965.

[18] Schaber, G. G., Eggleton, R. E., and Thompson, T. W., "Lunar Radar Mapping: Correlation Between Radar Reflectivity and Stratigraphy in Northwestern Mare Imbrium," Nature, Vol. 226, 1970, pp. 1236-1239.

[19] Thompson, T. W., et al., "Comparison of Geologic, Infrared, and Radar Mapping of Lunar Craters," Contribution 16 of the Lunar Science Institute, Houston, Texas, December 1970.

[20] Shorthill, R. W., and Saari, J. M., "Non-uniform Cooling of the Eclipsed Moon: A Listing of Thirty Prominent Anomalies," Science, Vol. 150, 1965, pp. 210-212.

[21] Lincoln Laboratory, "Radar Studies of the Moon," Final Report, Vol. 1, Lincoln Laboratory, Lexington, Mass., 1967.

[22] Thompson, T. W., "Radar Studies of the Lunar Surface Emphasizing Factors Related to Selection of Landing Sites," Research Report RS-73, Center for Radiophysics and Space Research, Cornell University, Ithaca, N. Y., April 1968.

SECTION 2.   IN SITU SURFACE MEASUREMENTS

LUNAR THERMAL ASPECTS FROM SURVEYOR DATA

Leonard D. Stimpson[*] and John W. Lucas[†]

Jet Propulsion Laboratory, Pasadena, Calif.

Nomenclature

A = bolometric or total solar albedo, dimensionless

A_c = area of cross section, m^2

A_i = area of ith surface, m^2

c = specific heat, cal/g-°K

C = heat capacity coefficient, w-sec/m^2-°K

F_{ij} = view factor from surface i to surface j

$$= \frac{1}{A_i} \iint \frac{\cos \omega_i \, \cos \omega_j \, dA_i \, dA_j}{\pi r^2}, \text{ dimensionless}$$

This paper presents the results of one phase of research carried out at the Jet Propulsion Laboratory, California Institute of Technology, under Contract No. NAS 7-100 sponsored by NASA.

A Lunar Thermal Properties Working Group was established by the Surveyor Project approximately 6 months before Surveyor I was launched. This Group continued its advisory activities until approximately a year and a half after Surveyor VII was launched. Composition of the Group was as follows: John W. Lucas, Chairman, James E. Conel, and William A. Hagemeyer, Jet Propulsion Laboratory; Robert Garipay, Hughes Aircraft Company; David Greenshields, NASA Manned Spacecraft Center; Hector C. Ingrao, DOT Transportation Research Center; Billy P. Jones, NASA Marshall Space Flight Center; Jack Saari, Boeing Scientific Research Laboratories.

[*]Member of Technical Staff.
[†]Manager for Research and Planetary Quarantine.

k   = thermal conductivity, w/m-°K

K   = kA_c/LA_i = conductivity coefficient, w/m^2-°K

L   = heat conduction length, m

$\dot{q}$ = conduction heat flux from inside of the compartment to the outboard face = 3.5 w/m^2

r   = distance between surfaces used in view factor, m

S   = solar radiation, w/m^2

T   = temperature, °K

α_s = solar absorptance

β = angle between direction of sun and normal to panel

γ = $(k\rho c)^{-1/2}$, thermal parameter, cm^2-sec$^{1/2}$-°K/g-cal

ϵ = emittance

ρ = density, g/cm^3

σ = Stefan-Boltzmann constant = 5.675 x 10^{-8} w/m^2-°K^4

ψ = sun elevation angle above lunar horizon, deg

ω = angle from surface normal used for view factor, deg

Subscripts

0   = compartment surface containing thermal sensor (inorganic white); α_{Os} = 0.20, ϵ_0 = 0.87

1   = sunlit lunar surface; ϵ_1 = 1 (brightness assumption)

1' = shaded lunar surface

2,3 = vertical sides of compartment (inorganic white)

4   = inboard surface (polished aluminum); α_{4s} = 0.10, ϵ_4 = 0.04

5   = bottom (polished aluminum)

6   = top (inorganic white) (exclusive of thermally isolated mirrors)

6' = compartment mirror surfaces; $\epsilon_{6'}$ = 0.79

7 = sunlit solar panel sides; ϵ_7 = 0.8 (solar cells)

7' = shaded solar panel sides; $\epsilon_{7'}$ = 0.84 (organic white)

8 = planar array antenna; ϵ_8 = 0.88 (black)

9 = solar stepping motor; ϵ_9 = 0.86 (organic white, 3M)

L = Lambertian

I.   Introduction

Five unmanned Surveyor spacecraft successfully landed on the moon and transmitted data back to earth. The Surveyor landing sites are described in Table 1, which lists the selenographic location, time of landing, and other information for each site; Fig. 1 shows the landed orientations of the spacecraft. The local terrain upon which each of the spacecraft landed was different. Surveyor I landed on a relatively smooth, nearly level surface encircled by hills and low mountains. Surveyor II failed during a midcourse maneuver. Surveyor III landed halfway down a 12-1/2° slope of a crater about 200 m in diameter and 15 m deep. Surveyor III also was visited by the Apollo 12 astronauts in November 1969. Surveyor IV ceased to transmit data just before landing near the subsequent Surveyor VI site. Surveyor V landed with one leg on the rimless edge of a 9 × 12-m crater 1.5 m deep, with the other two legs inside the crater, and was tilted about 20° from the lunar horizontal. Surveyor VI landed on a relatively smooth, flat surface. The local slope was less than 1°; after the hop made by the spacecraft, the local slope on the new site was about 4°. Surveyor VII landed in extremely rough terrain but with a local slope of only about 3°.

The behavior of the different spacecraft on the lunar surface varied. Surveyor I gave excellent data for two successive lunar days (June and July 1966), and partial data were obtained as late as the sixth lunar day. The spacecraft operated for 48 hr into the first lunar night. Surveyor III landed with the vernier propulsion system still at a thrust level almost equal to the lunar weight. After initial touchdown, it touched down on the crater slope two more times before coming to rest. On the second touchdown, all analog telemetry signals became erroneous. It was found that most of the analog data obtained in the lowest rate mode (17.2 bps) were fairly reliable and could be corrected with simple calibration factors. However, the over-all accuracy of telemetered temperatures from Surveyor III was estimated at ±6°K compared with that of ±4°K for

Table 1 Positional characteristics of Surveyor spacecraft

| Spacecraft | Selenographic coordinates, Atlas/ACIC System | | Selenographic location | Touchdown time | | Sun elevation above eastern horizon at touchdown, deg | Approximate local slope, deg |
	Latitude	Longitude		GMT, Date	GMT, hr:min:sec		
Surveyor I	2.46°S	43.23°W	Southwest part of Oceanus Procellarum (Ocean of Storms)	June 2, 1966	06:17:36	28.5	0.5
Surveyor III	2.99°S	23.34°W	Southeast part of Oceanus Procellarum	Apr. 20, 1967	00:04:17[a]	11.8	12.5
Surveyor V	1.4°N[b]	23.2°E[b]	Mare Tranquillitatis (Sea of Tranquility)	Sept. 11, 1967	00:46:42	16.4	19.5
Surveyor VI	0.51°N	1.39°W	Sinus Medii (Central Bay)	Nov. 10, 1967	01:01:04	2.8	1[c]
Surveyor VII	40.88°S	11.45°W	Ejecta blanket of Crater Tycho	Jan. 10, 1968	01:05:36	12.5	3

[a] Initial touchdown; second touchdown was at 00:04:41 GMT; final touchdown was at 00:04:53 GMT.
[b] Approximate.
[c] Before the hop. After the hop, the slope was about 4°.

SURVEYOR I (2.46°S, 43.23°W)
MARIA: LEVEL

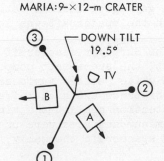

SURVEYOR III (2.99°S, 23.34°W)
MARIA: 200-m CRATER

SURVEYOR V (1.4°N, 23.2°E)
MARIA: 9-×12-m CRATER

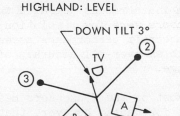

SURVEYOR VI (PRE-HOP) (0.51°N, 1.39°W)
MARIA: LEVEL

SURVEYOR VII (40.88°S, 11.45°W)
HIGHLAND: LEVEL

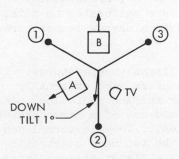

Fig. 1 Surveyor spacecraft landed orientations.

the other spacecraft. A solar eclipse by the earth during the
first lunar day (Apr. 24, 1967) offered the first opportunity
to observe such an event from the moon. Surveyor III shut
down 2 hr after sunset on the first lunar day.

Surveyor V, which operated for about 115 hr into the first
lunar night, also underwent a solar eclipse on the second
lunar day (Oct. 18, 1967) and operated for about 215 hr into
the second night. It operated for a short period of time during the fourth day, transmitting 200-line television pictures.
The vernier rocket engines on Surveyor VI were fired on the
lunar surface during the first day (Nov. 17, 1967), causing the
spacecraft to lift off from the surface and hop 2.4 m. Surveyor VI operated for about 40 hr into the night; it was
revived on the second day but gave thermal data for only a
short time. Surveyor VII, which operated for about 80 hr into
the first night, was successfully revived on the second day
and gave good thermal data during the day; however, contact
with the spacecraft was lost before sunset on the second day.

The Surveyor spacecraft is shown in Fig. 2. For the most
part, the spacecraft components in the sun-illuminated areas
had white painted surfaces that provided a low-solar-
absorptance and high-infrared-emittance thermal finish for
protection against the high midday solar intensity. Polished
aluminum surfaces were used on most of the underside to isolate
the spacecraft thermally from the lunar surface.

The temperature data of various points in the spacecraft
were provided by platinum resistance thermal sensors. Each
sensor was calibrated individually to $\pm 2^{\circ}K$; other nominal system inaccuracies degraded the over-all accuracy to $\pm 4^{\circ}K$. These
sensors were low resolution; other sensors, critical for
spacecraft performance assessment, were calibrated to $\pm 1^{\circ}K$,
with an over-all accuracy of $\pm 3^{\circ}K$ over a narrow temperature
range.

Most of the 75 sensors measured internal temperatures. It
was fortunate, however, that some sensors were externally
located and were responsive to lunar surface radiation, since
project funding and schedule prevented use of a more sophisti-
cated temperature measuring instrument. These external sensors were on the outside panels of the two main electronic
components, on the solar panel, on the planar array antenna,
and on the upper part of the mast.

Compartments A and B housed the spacecraft electronics and
battery. A thermal blanket of multilayer insulation surrounded
the components in each compartment, which, in turn, was covered

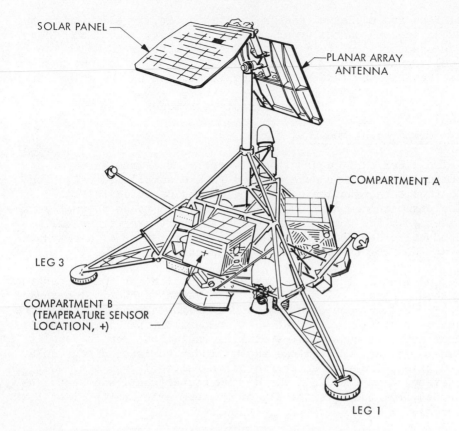

SOLAR PANEL

PLANAR ARRAY
ANTENNA

COMPARTMENT A

LEG 3

COMPARTMENT B
(TEMPERATURE SENSOR
LOCATION, +)

LEG 1

Fig. 2 Surveyor spacecraft configuration.

with outer aluminum panels. Heat was rejected from the
compartments during the hot lunar day through bimetal-actuated
thermal switches connected to highly polished Vycor mirrors on
the top of each compartment. At sunset, the thermal switches
opened to isolate the interior of the compartments from the
cold surroundings. Internal electrical heaters were available
to warm the compartments.

A thermal sensor was bonded to the polished-aluminum inner
surface of the outboard panel, i.e., to the aluminum surface
facing the blanket of each compartment (Fig. 2). Since the
outboard panels of the compartments had a strong radiative
coupling to the lunar surface but were virtually shielded from
the view of other spacecraft components, an analysis of lunar
surface brightness temperatures was possible.

The solar panel and planar array antenna shown in Fig. 2
were relatively low-heat-capacity planar surfaces. Each

surface had a thermal sensor located on it, in addition to two other sensors on the upper part of the mast. Data from these four sensors have been used in the model referred to as the "solar panel model." The most sensitive of these four sensors to the lunar surface temperature is located on the solar panel.

II. Analytical Techniques

A. Compartment Model

The lunar surface brightness temperatures from Surveyor thermal sensor data were derived from a simplified radiation heat balance equation based on elements depicted in Fig. 3.[1] The outboard compartment panel was assumed to be thermally isolated from the remainder of the spacecraft except for small conductive heat contributions from the interior of a compartment and from the sides of the compartment.

Equation (1) is the heat balance equation for the compartment outboard panel, which contains the thermal sensor. The terms in Eq. (1) in sequence from left to right are as follows. The energy radiated by the outboard compartment panel is equal to nine energy inputs: infrared (IR) radiation from the sunlit lunar surface, IR from the shaded lunar surface, direct solar radiation, indirect solar radiation (albedo) from the lunar surface, heat flux from inside the compartment, and heat conducted from the other panels of the compartment (the last four terms):

$$\epsilon_0 \sigma T_0^4 = \epsilon_0 \epsilon_1 F_{01} \sigma T_1^4 + \epsilon_0 \epsilon_1 F_{01'} \sigma T_{1'}^4 + \alpha_{0s} S(\cos \beta + AF_{01} \sin \psi) +$$

$$\dot{q} + K_{20}(T_2 - T_0) + K_{30}(T_3 - T_0) +$$

$$K_{50}(T_5 - T_0) + K_{60}(T_6 - T_0) \tag{1}$$

The equations for the remaining five panels of the compartment are similar in form to Eq. (1), where the index 0 is replaced sequentially by 2 through 6 and conduction terms are adjusted accordingly.

An assumed value of 200°K (-100°F) is used for the shaded lunar surface. This assumed value is adequate, since it is near the actual value and the shadow heat input term was only of minor significance. This was due to small view factor values involved and then only during a small part of the lunar day. An approximate value of 3.5 w/m^2 is used for the heat flux $\dot{q}$ from the inside; this was obtained from test in vacuum

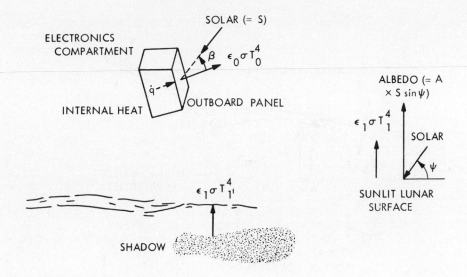

Fig. 3   Heat transfer for an outboard compartment face.

(liquid N_2-cooled walls and room-temperature compartment).
The heat conduction term was important after sunset and during
the totality of the eclipse.

B.   Solar Panel Model

A predominantly radiative heat-balance equation was used for
determining the lunar surface temperatures from the Surveyor
solar panel depicted in Fig. 4.   The model was used only for
postsunset and eclipse (umbra) periods, since it was insensitive to the lunar surface during the daytime because of the
variable electrical power loading and the high solar insolation.   The thermal sensor was centrally located on the solar
panel.   Both sides of the solar panel emitted heat [left side
of Eq. (2)] and also viewed [right side of Eq. (2)] the lunar
surface, the planar array, the white tops of the compartments,
and the mirror surfaces on top of the compartments.   Also
included on the right side is the heat conduction through the
mast and the heat capacity of the solar panel:

$$(\epsilon_7 + \epsilon_{7'})\sigma T_7^4 = (\epsilon_7 F_{71'} + \epsilon_{7'}F_{7'1'})\epsilon_{1'}\sigma T_{1'}^4 +$$

$$(\epsilon_7 F_{78} + \epsilon_{7'}F_{7'8})\epsilon_8\sigma T_8^4 + \epsilon_{7'}\epsilon_6 F_{7'6}\sigma T_6^4 +$$

$$\epsilon_{7'}\epsilon_{6'}F_{7'6'}\sigma T_{6'}^4 + K_{97}(T_9 - T_7) - C_7\frac{dT_7}{dt} \quad (2)$$

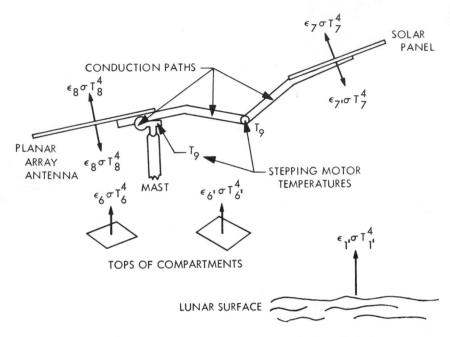

Fig. 4   Solar panel heat transfer model.

A similar equation exists for the planar array antenna, where indices 7 and 8 are interchanged.

Certain coordinate transformations were required to determine the view factors and sun angles for each of the compartment panels, the solar panel, and the planar-array antenna. Each view factor required a vectorial description of the normal to the panel and to the local lunar surface. These coordinate transformations and vectors are given by Stimpson et al.[1]

C.   Error Analysis

An error analysis was performed on the Surveyor-based data because of discrepancies found among the results obtained from compartment, solar panel, and earth-based measurements. The main error contributors for the compartment panels were found to be the inaccuracy in view factor F_{01}, the uncertainty in thermally sensed temperature T_0, and the uncertainty in internal heat loss $\dot{q}$ after sunset. Others that contributed to a lesser extent were α_{0s}, ϵ_0, S, and β. A similar study of the solar panel, Eq. (2), also showed the view factors and the telemetered temperature data to be the principal error contributors for the solar panel source.

Heat conducted around the outside of the compartments from
the other panels creates an upward bias in the lunar surface
temperature. This is mainly due to the two vertical side faces
viewing more warm lunar surface than the outboard face, which
was tilted back 20°. Note also that one of the sides is almost
always illuminated by the sun during the day. In some missions, either early in the morning or late in the afternoon,
the sun also heated the inboard face to rather high temperatures, since it was a polished aluminum (low-emissivity) surface. These effects have been included in the analysis.

Figure 5 shows the results obtained from the error analysis
performed on the Surveyor V data. Figure 5a shows the individual contributing error sources from compartment A data and
Fig. 5b from compartment B. These are reflected in subsequent
figures as error bands placed about the results obtained.
Note that the errors become magnified near sunset. The individual errors were assumed to be independent from each other;
the total errors were obtained by taking a root-sum-square of
the individual errors. Measurements from compartments and
solar panel are not independent.

Error sources from the Surveyor V solar panel postsunset data
are shown in Fig. 5c. The total error is similar to that
obtained from the compartments. The temperature sensor error
is the predominant source in the three figures, especially
after sunset. The mast conduction error shown in Fig. 5c is
primarily due to uncertainties in temperature sensor measurements on the mast.

The individually assumed errors for this analysis were

$$\Delta T_i = \pm 4°C \qquad\qquad \Delta F_{ij} = \pm 10\%$$

$$\Delta \epsilon_i = \pm 0.02 \text{ for painted surfaces, } +0.02, -0.01 \text{ for polished} \\ \text{aluminum}$$

$$\Delta \alpha_i = \pm 0.02 \qquad\qquad \Delta \dot{q} = \pm 20\% \qquad\qquad \Delta K_{ij} = \pm 10\%$$

III. Lunar Surface Temperatures

A. Earth-Based Data

Since results from Surveyor are presented and compared with
earth-based data in this chapter, the latter are briefly
discussed first (also see Chapter 1a). Throughout this
chapter, brightness temperature is used in the usual sense,

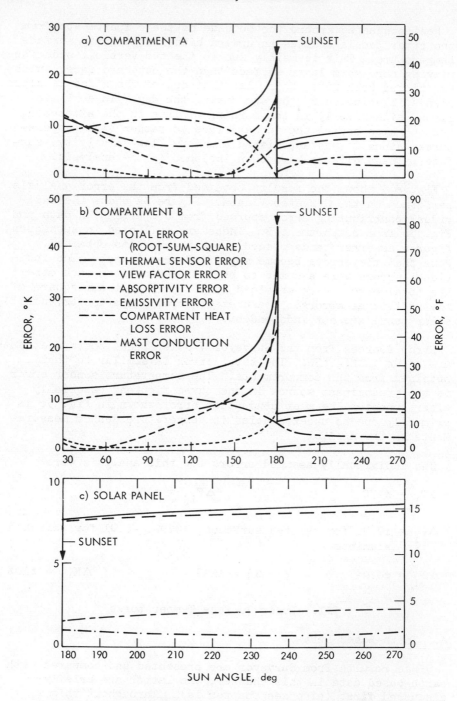

Fig. 5   Errors from Surveyor V.   a) compartment A;
b) compartment B; c) solar panel.

that is, the experimentally observed temperature a surface with unit emissivity must have to produce the measured response.

1. Thermophysical properties of Surveyor sites. The thermophysical properties can be determined from postsunset, eclipse, or lunation cooling curves. The most extensive eclipse measurements are those of Saari and Shorthill[2,3] (made during the Dec. 19, 1964 eclipse). Data on isotherms during totality for the equatorial region have been published[4]; the measurements revealed anomalous cooling of features of a wide range of sizes, varying from kilometer-sized craters to the entire maria. It would not, therefore, be surprising if thermal heterogeneity were found to dimensions much smaller than measured by the earth-based eclipse measurements. It should be noted that the resolution of earth-based data is 18 km, whereas that from the Surveyor compartments is 18 m and from the solar panel about 50 m.

The area in which Surveyor I landed is one with small horizontal thermal gradients; thus, it contains the highly insulating properties that typify the general lunar surface. The Surveyor III, V, and VI regions also appear to be relatively bland and at the limit of resolution of the earth-based measurements.

The crater Tycho is an outstanding thermal anomaly on the lunar surface from the standpoint of the temperature difference over its environs and the size of the area affected. The Surveyor VII site is within the anomalous area surrounding the crater.

2. Lunation temperatures. Lunation calculations[5] of the homogeneous model are used assuming constant thermophysical properties. These properties are characterized by the thermal parameter $\gamma = (k\rho c) \exp(-1/2)$. This constant model[6], however, cannot adequately represent the earth-based measurements during both eclipse and postsunset, since, at the least, they require different constant values of γ. During illumination, the model predicts temperatures essentially in agreement with Lambertian [Eq. (3) below] when γ is greater than 500. A particulate model of the lunar soil has been proposed[7] which agrees with both the earth-based eclipse and postsunset cooling. Winter and Saari,[7] and Jones[8] take into account soil property variations with temperature and depth of soil.

A γ value[5] of 800 is typical for the lunation of the equatorial Surveyor sites and was derived from earth-based postsunset measurements of mare areas in the eastern section.[9] The larger γ values given in the following paragraphs resulted from earth-based eclipse measurements. The difference in γ is thought to be a consequence of heat exchange from only the uppermost millimeters of soil during an eclipse, whereas a different type of soil at a lower depth is involved during the lunation warming and cooling phases. The depth of penetration of the thermal wave is an order of magnitude greater during lunation than eclipse, since it is proportional to the square root of the period of variation.

3. <u>Lambertian temperatures</u>. The calculated lunar surface Lambertian temperatures[5] for the homogeneous model at the Surveyor I landing site are shown in Fig. 6. The Lambertian temperature T_L (with unit surface emissivity assumed) is defined by the expression

$$\sigma T_L^4 = (1 - A)S \sin \psi \tag{3}$$

The Lambertian temperature is that which a perfectly diffuse surface would have in order to radiate the same energy as is absorbed. To calculate T_L, the bolometric albedo, i.e., the reflectance of the lunar surface to the total solar spectrum, of each site must be known. For this purpose, the simultaneous infrared and photometric scan data of Saari and Shorthill[10] were used. Implicit in Eq. (3) is that the heat capacity of the lunar surface material is zero.

The specific values for solar constant and bolometric albedo used for each mission are given in Table 2, based on a mean value of 1390 w/m^2 (442 Btu/hr-ft^2), where the solar distance variation is from the American Ephemeris and Nautical Almanac[11]. The time scale was fixed assuming a flat moon surface at sunset. The γ = 800 intermediate curve in Fig. 6 is considered most representative of the site. The lunar surface Lambertian temperatures for Surveyors III, V, and VI are nearly identical to that shown in Fig. 6 for Surveyor I.

The calculated Lambertian temperatures and earth-based temperatures at the Surveyor VI landing site are shown in Fig. 7. These earth-based measurements show the directionality of lunar infrared emission (also see Chapter 1a); near local noon, when the surface was observed from the same general direction as the sun, the measured temperatures were higher than the calculated Lambertian temperatures. Earth-based <u>eclipse</u> observations show cooling during totality comparable to that for a homogeneous model with a γ of 1100.

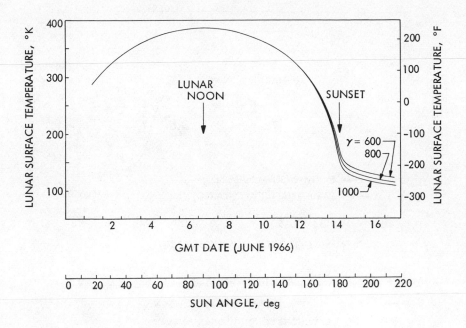

Fig. 6 Calculated Lambertian temperature for Surveyor I site.

Table 2   Solar constant and view factors

Surveyor	Time	S, w/m^2	Bolometric albedo, A	F_{12} Comp. A	Comp. B
I	Landing	1352	0.052	0.28	0.29
III	Landing	1386	0.076	0.31	0.41
V	Landing	1375	0.077	0.25	0.26
V	Eclipse	1408	0.077	0.23	0.27
VI	Landing	1423	0.084	0.32	0.32
VI	After hop	1430	0.084	0.35	0.32
VII	Landing	1442	0.17	0.34	0.33

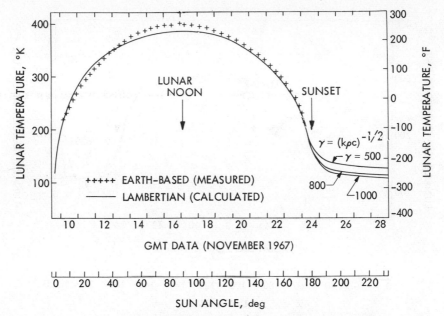

Fig. 7   Earth-based and calculated brightness
temperatures for Surveyor VI site.

The calculated Lambertian temperatures for the Surveyor VII
landing site are shown in Fig. 8. Also presented are the
earth-based measured temperatures, which again show a directional effect distributed over a larger portion of the lunar
day. During the Dec. 19, 1964 eclipse, Ingrao et al.[12] made
measurements of Tycho to a 9-arc-sec resolution up to a few
minutes before the end of totality. These <u>eclipse</u> observational data fit the cooling curve for a homogeneous model, with
γ = 450 inside the crater and with γ = 1100 in the environs
outside the crater.

Although no earth-based measurements of the local Surveyor
VII region were made during the lunar <u>night</u>, it was possible
to obtain a postsunset cooling curve by interpolating between
the center and the environs, resulting in γ = 550 at the site.

4. <u>Eclipse temperatures</u>. Figure 9 shows a cooling curve for
the Surveyor III site from earth-based measurements obtained
during the December 19, 1964 eclipse.[2] When this curve was
compared with the theoretical eclipse cooling curves for a
homogeneous model,[6] it was possible to infer a value for γ of
1400. Values of γ in this range, as determined from eclipse
calculations, are representative of the insulating material
that characterizes much of the lunar surface. The warming
curve in Fig. 9 represents calculated equilibrium surface
temperatures corresponding to the insolation at each time.

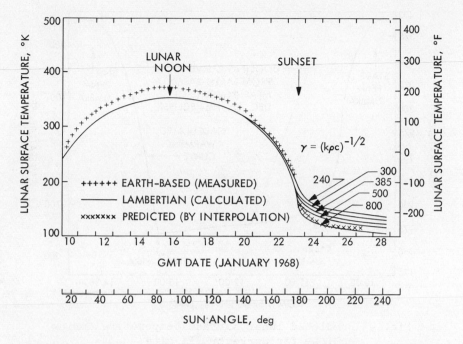

Fig. 8 Earth-based, calculated, and predicted brightness temperatures for Surveyor VII site.

Figure 10 is an eclipse cooling curve for the Surveyor V site from earth-based measurements. By using the theoretical eclipse cooling curves for a homogeneous model,[6] a γ of 1350 was obtained for the lunar surface material.

5. Directional effects. It has been determined that, when the lunar surface is illuminated by the sun, the observed brightness temperature is not constant for different angles of observation; i.e., the surface does not behave like a Lambertian surface. This effect, ascribed to surface roughness, causes the brightness temperature to be higher when the sun/surface/observer angle is small (see Chapter 3c).

To correct the earth-based measurements for directional effects, measurements over the entire lunar disk at three sun angles were used. For a sun elevation angle of 90°, the measurements of Sinton[13] were taken and show the variation in radiance from the subsolar point as a function of the angle of observation. For two other sun angles of 30° and 60°, the infrared scan data for different phases made by Shorthill and Saari[14] were used. The albedo corrections for each point were

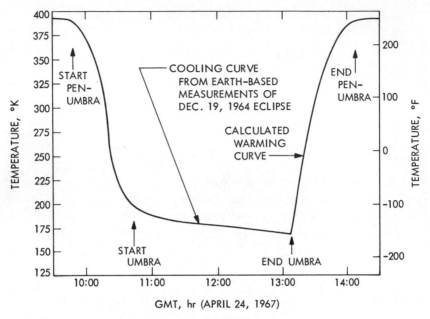

Fig. 9   Predicted lunar surface temperature during eclipse for Surveyor III site.

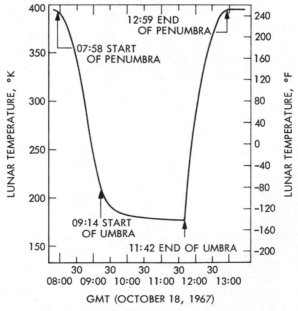

Fig. 10   Eclipse brightness temperature for Surveyor V site from earth-based data.

made from their full-moon photometric data. The directional
factor was determined using a calculated Lambertian temperature
at each point.

Directional factors were obtained from global measurements
made on a variety of features. It is possible, therefore, for
a small area such as a Surveyor landing site to have directional effects different from the average surface if the local
roughness or surface configuration differed significantly from
the average.

B. Surveyor Results

Spacecraft temperatures telemetered back were subjected to
analysis with equations typified by Eqs. (1) and (2). ϵ_1 was
taken to be unity so that lunar surface brightness temperatures
were obtained.

1. <u>Daytime</u>. The complete analysis described in Sec. IIC was
specifically applied to the Surveyor V daytime data. The lunar
surface daytime temperatures obtained from compartment data are
compared in Fig. 11 with the Lambertian prediction and earth-
based measurements. Directionality during the midday on the
earth-based results has been mentioned before. There also is
some evidence of directionality in the Surveyor results, particularly in Fig. 11b. The outboard face of compartment B
viewed the west, so that a directional effect (temperature
increase) would be expected during the lunar morning, as is
shown. The error band in the morning for compartment B suggests rather clearly that a directional trend exists. The
trend is not as clear in Fig. 11a for compartment A, which
viewed the southeast rather than directly east or west and
thus would not be expected to show a strong directionality
effect in the lunar afternoon.

It is expected that similar results would be obtained if the
complete analysis were applied to data from the other
spacecraft.

2. <u>Postsunset</u>. The postsunset lunar surface temperatures
derived from compartment data are presented in Figs. 12-15
for all the missions, except for that of Surveyor III, which
transmitted only 2 hr after sunset. Temperatures derived from
Surveyor V postsunset data during the second lunar night also
are included, since the transmission lasted the longest of any
mission (9 earth days). Temperatures are also presented from

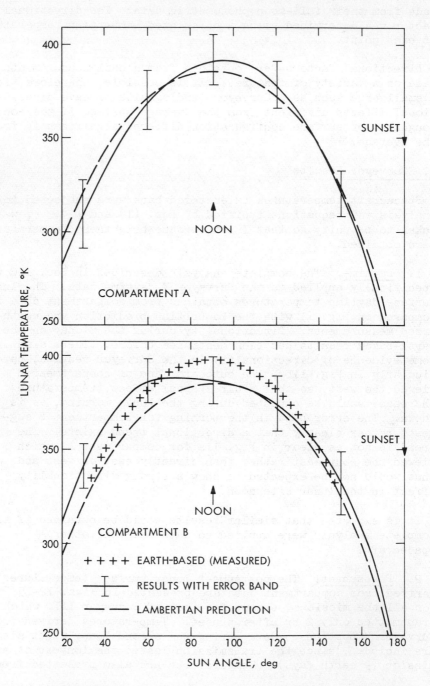

Fig. 11   Daytime lunar surface temperatures from Surveyor V.
a) compartment A; b) compartment B.

solar panel postsunset data, except for Surveyor VII, where the
solar panel thermal sensor failed at sunset. In addition, these
results are compared with earth-based measurements and γ curves.
The γ curves assume a lunar surface having thermal characteristics that are temperature-independent and homogeneous. Actually, the lunar surface properties are expected to vary with
depth of soil and with temperature.[7,8] Thus a difference in
behavior between the γ curves and the compartment band shortly
after sunset is to be expected.

The postsunset lunar surface temperatures from the Surveyor I
compartment and solar panel data are shown in Fig. 12. The
earth-based measurements from Wildey, Murray, and Westphal[9]
fall between compartment and solar panel results. Error bands
were not calculated for Surveyor I data, because of insufficient
funds, but would be similar to those shown in Fig. 13.

The Surveyor V compartment data, shown in Fig. 13, are
obtained from four compartment sources (both compartments on two
successive nights). They fall close to each other, with an
average γ of about 750 early in the night to about 600 later
into the night. The Surveyor V solar panel data from the two
nights also fall close to each other, with an early night γ near
1200 and later about 1000. Data from the other Surveyor spacecraft only reach the early night condition. Error bands are
also included (from Fig. 5) which indicate some overlap of the
data.

The Surveyor VI compartment data are shown in Fig. 14. The
compartment, solar panel, and earth-based measurements are all
in close agreement. Error bands were not calculated but would
be similar to those in Fig. 13.

The cooling curves for the two compartments on Surveyor VII
are shown in Fig. 15 together with curves for constant γ. (As
noted previously, no postsunset solar panel data exist.) The
compartment curves are quite different from each other and also
from mare data. These differences are attributed to a more
rocky terrain, especially for compartment B (see Figs. 16 and
17). Recall that Surveyor VII landed in a highlands region
near Tycho.

Returning to Fig. 15, note that the effective gammas from the
compartment data are very low shortly after sunset. The values
later in the night are to be compared with an earth-based estimate of 550 for the landing site, as mentioned in Sec. IIIA3.

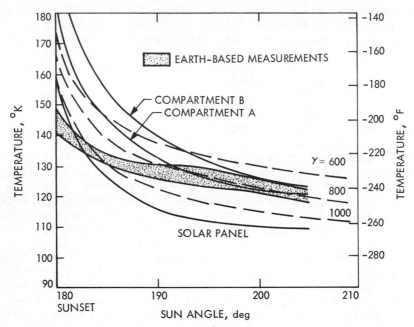

Fig. 12   Postsunset lunar surface temperatures from
Surveyor I.

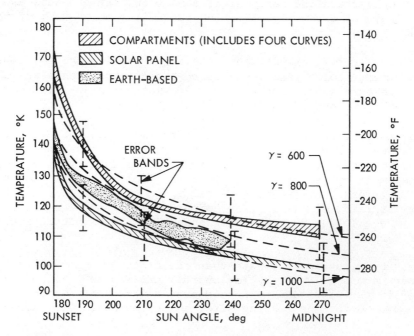

Fig. 13   Postsunset lunar surface temperatures from
Surveyor V.

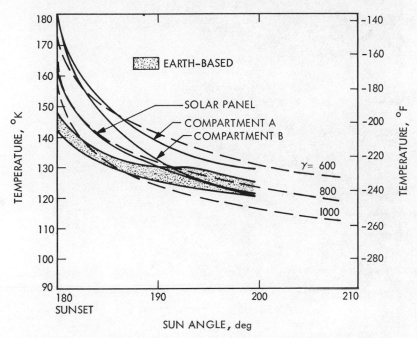

Fig. 14   Postsunset lunar surface temperatures from
          Surveyor VI.

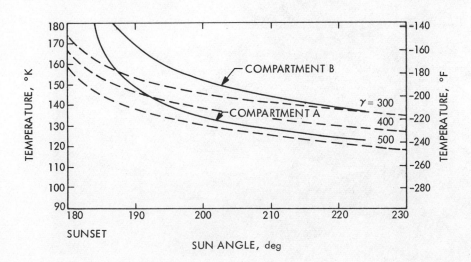

Fig. 15   Postsunset lunar surface temperatures from
          Surveyor VII compartment data.

Fig. 16 Lunar scene viewed by compartment A of Surveyor VII.

Fig. 17 Lunar scene viewed by compartment B of Surveyor VII.

Figure 18 consolidates postsunset results from Surveyors I, V, and VI, which landed in maria. The compartments result in higher lunar surface postsunset temperatures (and lower γ values) that do the solar panel data, whereas the earth-based data fall into the intermediate region. Figure 15 shows that the temperatures measured by the Surveyor VII compartments are considerably greater in the highlands near Tycho than the earth-based measurements, which are more representative of maria. Saari[15] estimated an earth-based γ for this highland region of 550 by interpolation.

A summary of the γ values is given in Table 3 for postsunset telemetry data from the Surveyors and earth-based measurements.

Table 3   Postsunset gamma comparisons

Surveyor	From compartments	From solar panel	Earth-based
I	750	1200	850
V	600 - 750	1000 - 1200	850
VI	750	850	850
VII	450 (A)		550
	300 (B)		

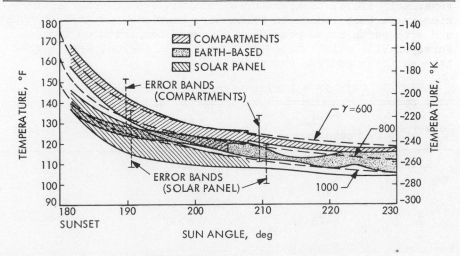

Fig. 18   Postsunset lunar surface temperatures from Surveyors I, V, VI, and earth-based data.

3. <u>Eclipse</u>. The sun was eclipsed by the earth at the Surveyor III site on the first lunar day and at the Surveyor V site on the second lunar day. An analysis of the eclipse data as well as the Surveyor V postsunset data was reported earlier.[16] Especially note the crater detail shown in Fig. 5 of Ref. 16 obtained from TV pictures.

Surveyor III lunar surface temperatures from the compartment and solar panel data are shown in Fig. 19. The compartments predict higher temperatures as they did after sunset. It is seen that the temperatures derived from the solar panel data and the earth-based measurements cross the constant curves; thus the concept of a constant γ is not in agreement with the measured results. (The solar panel result is averaged over the umbra phase, whereas the comparison is made at the start of umbra for earth-based measurements.) Error bands were not calculated for Surveyor III, but they would be similar to those of Surveyor V, shown in Fig. 20.

The lunar surface temperatures from the Surveyor V data during the eclipse are compared in Fig. 20. Error bands are shown as well for temperatures derived from the compartments and solar panel late in the umbra phase. Values of gamma from the different sources are given in Table 4.

IV. Conclusions

The spatial resolution of lunar surface temperatures from Surveyor data was 1000 times better than that of temperatures from earth-based measurements; yet the correspondence is remarkably close. Temperatures obtained from the Surveyor spacecraft that landed in maria were consistent with each other and with earth-based results. However, temperatures from Surveyor VII, which landed in a highlands region, were higher than anticipated from earth-based data.

Table 4   Eclipse gamma comparisons

Surveyor	From compartments	From solar panel	Earth-based
III	500 (A)	1000	1420
	400 (B)		
V	425 (A)	1000	1350
	385 (B)		

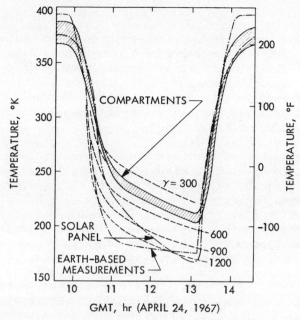

Fig. 19   Eclipse lunar surface temperatures from Surveyor III.

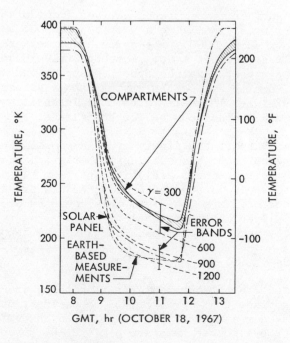

Fig. 20   Eclipse lunar surface temperatures from Surveyor V.

A. Maria

The daytime lunar surface temperatures (uncorrected for directionality) derived from Surveyor V compartment data show the
presence of directionality, in agreement with earth-based measurements. It is believed that the directional trends from
other spacecraft would exist to a similar degree. Otherwise,
compartment data and earth-based measurements made during the
day corresponded well.

The postsunset lunar surface temperatures derived from
Surveyor compartment data give a γ varying from about 750 early
in the night to 600 later in the night. The corresponding temperatures from Surveyor solar panel data give a γ of about 1200
(early) to 1000 (later). The earth-based postsunset measurements are bracketed by these results with an average γ equal to
850. The most probable postsunset temperature measurements
correspond closely to the earth-based measurements, which demon-
strate that the lunar soil is a finely particulate, highly insulating material. Further, the fact that Surveyor results from
a very small sampling of the lunar surface agree with earth-
based data indicates that lunar thermophysical properties are
very uniform across the surface on the scale of tens of meters.

The lunar surface temperatures derived from Surveyor compartment data during eclipse give a γ of 385 to 500, and from the
solar panel of about 1000. The earth-based measurements give
higher γ values of about 1400. Note in Tables 3 and 4 that
Surveyor compartment γ values from postsunset and eclipse data
differ considerably, as do the earth-based γ values, whereas
the solar panel results seem to be more insensitive; the latter
is probably due to unaccounted for heat leaks.

The results support the finely granulated soil theory, but
they also suggest that the soil characteristics under and
adjacent to the spacecraft are different from the typical undisturbed surface. It is suggested that the disturbed surface
exhibits a more compact behavior; this could be due to the
upper fluffy layer having been blown away by the retrorocket
exhausts or by a physical compaction caused by the exhausts.

B. Highlands

Lunar surface temperatures higher than those from earth-based
data were obtained from Surveyor VII compartment data both during the lunar day and after sunset. Values for γ from compartments A and B postsunset data are 450 and 300, respectively, and
appreciably lower than the earth-based value of 550. These
lower values from Surveyor VII were apparently caused by the

presence of the rock-like debris seen in the TV pictures, particularly in the view of compartment B, Fig. 17.

C. General Comment

It should be noted that it is both highly desirable and well within the state-of-the-art of infrared sensors to include a steerable infrared sensor on future spacecraft planetary landers. The information that could be obtained would be superior to that described in this chapter.

References

[1] Stimpson, L. D., Hagemeyer, W. A., Lucas, J. W., Popinski, Z., and Saari, J. M.,; "Revised Lunar Surface Temperatures and Thermal Characteristics from Surveyor," in Analysis of Surveyor Data, TR 32-1443, Jet Propulsion Laboratory, Pasadena, Calif., June 30, 1969, pp. 1-33.

[2] Saari, J. M. and Shorthill, R. W., "Thermal Anomalies of the Totally Eclipsed Moon of December 19, 1964," Nature, Vol. 205, 1965, pp. 964-965.

[3] Shorthill, R. W. and Saari, J. M., "Non-uniform Cooling of the Eclipsed Moon: A Listing of Thirty Prominent Anomalies," Science, Vol. 150, No. 3693, 1965, pp. 210-212.

[4] Saari, J. M. and Shorthill, R. W., "Isotherms in the Equatorial Region of the Totally Eclipsed Moon," Doc. Dl-82-0530, Boeing Scientific Research Laboratories, Seattle, Wash., Apr. 1966.

[5] Jones, B. P., "Diurnal Lunar Temperatures," AIAA Paper No. 67-289, April 17-19, 1967.

[6] Jaeger, J. L., "Surface Temperature of the Moon," Australian Journal of Physics, Vol. 6, No. 10, 1953.

[7] Winter, D. F. and Saari, J. M., "A New Thermophysical Model of the Lunar Soil," Doc. Dl-82-0725, Boeing Scientific Research Laboratories, Seattle, Wash., 1968; also "A Particulate Thermophysical Model of the Lunar Soil," Astrophysical Journal, Vol. 156, No. 3, June 1969, pp. 1135-1151.

[8] Jones, B. P., "Density-Depth Model for the Lunar Outermost Layer," Journal of Geophysical Research, Vol. 73, No. 24, 1968, pp. 7631-7635.

[9]Wildey, R. L., Murray, B. C., and Westphal, J. A., "Reconnaissance of Infrared Emission From the Lunar Nighttime Surface," Journal of Geophysical Research, Vol. 72, No. 14, 1967, pp. 3743-3749.

[10]Saari, J. M., and Shorthill, R. W., "Isothermal and Isophotic Atlas of the Moon," Report CR-855, NASA, Sept. 1967.

[11]American Ephemeris and Nautical Almanac, for the Year 1965, U.S. Government Printing Office, Nautical Almanac Off., U.S. Naval Observatory, 1963.

[12]Ingrao, H. C., Young, A. T., and Linsky, J. L., "A Critical Analysis of Lunar Temperature Measurements in the Infrared," The Nature of the Lunar Surface, Proceedings of the 1965 IAU-NASA Symposium, Johns Hopkins Press, Baltimore, Md., 1966, pp. 185-211.

[13]Sinton, W. M., "Temperatures of the Lunar Surface," Physics and Astronomy of the Moon, edited by S. Kopal, Academic Press, New York, 1962, Chapter 11.

[14]Montgomery, C. G., Saari, J. M., Shorthill, R. W., and Six, N. F., "Directional Characteristics of Lunar Thermal Emission," Doc. D1-82-0568, Boeing Scientific Research Laboratories, Seattle, Wash., 1966; also AIAA Paper 67-291, Apr. 17-20, 1967.

[15]Lucas, J. W., Hagemeyer, W. A., Saari, J. M., Stimpson, L. D., and Vickers, J. M. F., "Lunar Surface Temperatures and Thermal Characteristics, Surveyor Project Final Report. Part II: Science Results," TR 32-1265, Jet Propulsion Laboratory, Pasadena, Calif., June 1, 1968.

[16]Vitkus, G., Lucas, J. W., and Saari, J. M., "Lunar Surface Thermal Characteristics During Eclipse from Surveyors III, V and After Sunset from Surveyor V," Progress in Astronautics and Aeronautics: Thermal Design Principles of Spacecraft and Entry Bodies, Vol. 21, Academic Press, New York, 1969, pp. 489-505.

LUNAR SURFACE TEMPERATURES
FROM APOLLO 11 DATA

Paul J. Hickson*

Bellcomm, Inc., Washington, D.C.

Introduction

The Early Apollo Science Experiments Package (EASEP) was
emplaced on the lunar surface by an Apollo 11 astronaut during
the first lunar landing in Mare Tranquilitatis. EASEP was a
solar-powered, condensed version of the standard nuclear-
powered Apollo Lunar Surface Experiments Package (ALSEP), to
be deployed as a multiexperiment observatory at each subsequent lunar landing site. In addition to a scientific experiment, a seismometer, EASEP contained an engineering experiment
(Dust, Thermal, and Radiation Engineering Measurement, DTREM I)
consisting of a set of six sensors to measure the lunar
environment.

One of the sensors, a precision nickel resistance thermometer of wide dynamic range, is mounted vertically and is
facing outboard of EASEP so that a good view of the lunar surface is obtained. Analysis of the nickel thermometer data obtained during the lunar daytime yields an average brightness
temperature of the lunar surface area viewed. Apollo 14 and 15
ALSEP's will carry an improved version, designated DTREM II,
which will yield lunar nighttime temperatures as well. DTREM
II will have an unobstructed view of the lunar surface and
carry an extra thermometer to measure the heat leaks from the
brightness thermometer.

In his recent review article on the infrared moon, Shorthill[1] comments that data on the nighttime temperature of the
moon are incomplete, since such data are difficult to obtain
from earth. The Surveyor spacecraft were equipped with bat-
teries as well as solar panels, but only limited in situ thermal data were obtained after sunset. The Apollo DTREM
measurements were conceived as a spatial extension of the

M. P. Odle assisted greatly in the computer programming.
*Physicist, Member of Technical Staff.

Surveyor results to other landing sites, from which ground-
truth in the form of return rock samples would be available,
and as a temporal extension to nighttime phase angles over
many complete lunations. The Apollo program permitted some
attention to thermometer errors; therefore, the experiment
design provides insulation not available to the Surveyor exper-
imentors, and an increase in brightness temperature accuracy
is expected.

Sketch of the Experiment

Figures 1 and 2 are sketches of the DTREM I package indicating the thermal isolation and thermal finish of the thermometer. Figure 3 shows the DTREM I location on EASEP, and Fig.
4 is a plot of the telemetered data from the first and second
lunar days.

The lunar surface brightness temperature is found from the
nickel thermometer energy-rate balance; i.e., the net rate of
energy flow out of the detector is equal to the net rate of
energy flow into the detector. We write this equation in the
form

$$\varepsilon_{Ni}\sigma T^4_{Ni} = \varepsilon_{Ni}F_{NiS}\,\varepsilon_S\sigma T^4_S + \dot{q}/A_{Ni} \qquad (1)$$

Where ε is the emissivity, σ is the Stefan-Boltzmann constant,
T is the absolute temperature, F the geometric view factor, Ni
stands for nickel resistance thermometer, and S stands for
lunar surface. The geometric view factor F_{SNi} is defined as
the fraction of the radiant energy emanating from finite surface area A_S which is intercepted by another finite surface
A_{Ni} and by reciprocity $A_SF_{SNi}=A_{Ni}F_{NiS}$. The term $\dot{q}/A_{Ni}$ is the
net algebraic sum, divided by the thermometer area, of all
heat leaks and heat inputs to the thermometer and of the rate
of internal energy storage in the thermometer. The heat leaks
and heat inputs to the thermometer include reflections and
emissions from all nearby surfaces of the EASEP spacecraft,
and conduction from the DTREM housing to the nickel thermometer. The internal energy storage is negligible and a steady
state analysis was possible.

The geometric view factors of the thermometer to the various
surfaces are listed in Table 1. Since T_{Ni} is measured and the
terms of $\dot{q}/A_{Ni}$ are measured or can be calculated from other
EASEP spacecraft housekeeping temperatures, the lunar surface
brightness temperature, defined as T_S for $\varepsilon_S = 1$, is found at
each sun angle from Eq. (1) with F_{NiS} and $\dot{q}$ adjusted to the
sun angle.

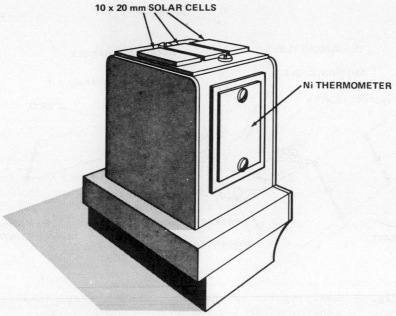

Fig. 1 Dust, thermal, and radiation engineering measurement package (DTREM I) as on EASEP/Apollo 11 and on ALSEP of Apollo 14 and 15.

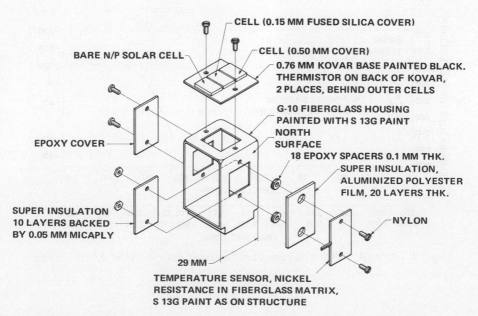

Fig. 2 DTREM I as on Apollo 11 EASEP, thermal details.

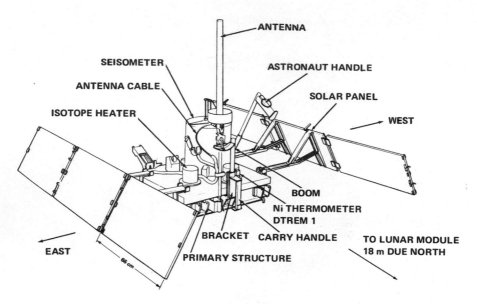

Fig. 3   Early Apollo science experiment package (EASEP)
         showing DTREM I geometry.

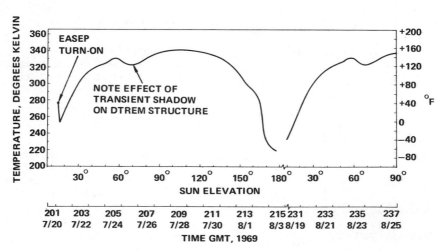

Fig. 4   DTREM I Ni thermometer temperature from EASEP data.

Table 1 Geometric view factors from the nickel thermometer to
various surfaces of EASEP[a]

Surface	Geometric view factor
Lunar surface	0.387^{b}
Black sky	$(0.312)^{c}$
Bracket, carry handle	0.160
Boom, astronaut handle	0.049
West, solar panel	0.025
Rock, east part	0.014
Rock, west part	0.025
LM descent stage	0.004
Laser experiment array	0.012
East solar panel	0.012
	1.000

[a]See Fig. 3

[b]Includes shadowed and unshadowed lunar surface. The sunlit
lunar surface has a geometric view factor, depending on sun
angle (Fig. 9), between 0.099 and 0.387. The geometric view
factor to the shadowed lunar surface is 0.387 diminished by
the view factor to the sunlit surface.

[c]An effective value for the black sky view factor of 0.418
was used, in agreement with Eq. (1), $F_{eff} = 1 - \Sigma \varepsilon_i F_i$.

The EASEP geometry is fairly complex, but careful thermal
analysis and detailed calculation of the appropriate corrections, yield, in principle, the following: 1) an average
equivalent brightness temperature of the unshadowed lunar
surface viewed; 2) an equivalent range of the surface thermal
parameter $(k\rho c)^{-\frac{1}{2}}$, where k is the thermal conductivity, ρ is
the mass density, and c is the heat capacity of the lunar sur-
face layer; and 3) the directional dependence of the lunar
surface thermal emission. Results of this type were reported
previously by Stimpson et al.[2] for all landed Surveyor
spacecraft.

Moreover, comparison of data from successive months may
yield the degradation rates of the thermal coatings used or
viewed. Note, however, that the EASEP thermometer is mounted
vertically, facing north, so that direct sun viewing is
avoided and coating degradation is expected to be small. Also,
the third objective was not met; i.e., the directional dependence of the lunar surface thermal emission could not be

determined from the EASEP because the thermometer faces north,
normal to the sun-to-surface direction, the very direction
that in previous measurements has been shown to be the important one.[3] The future ALSEP data should yield more complete
results.

Description of the Instrument

The thermometer used in the experiment is a nickel wire
resistor made by the Tyland Corporation. The resistor, which
has a nominal 5500-ohm resistance at the ice point, is connected to ground and also through a 15,000-ohm temperature-
insensitive precision dropping resistor to the +12 $\pm$ 1% regu-
lated EASEP power line as seen in Fig. 5. The thermometer
voltage drop is fed to the EASEP 8-bit analog-to-digital con-
verter and is measured once every 54 sec. The voltage of the
+12-v supply is measured 4 sec before each temperature is measured and so introduces only a small (<1/4%) measurement error.
The thermometer has a dynamic range of 84°K (-308°F) to 408°K
(274°F) and the ΔT, corresponding to 1/2 bit, is almost constant at 0.8°K over this entire range. According to the manu-
facturer, the calibration of the thermometer to an absolute
temperature scale is accurate to at least 0.8°K over the
entire range, since three calibration points, the steam point,
the ice point, and the carbon dioxide point, are taken for
each sensor. The thermometer and its circuit have good stabil-
ity chracteristics and drift less than 0.1°K/y.

For proper understanding of the thermal analysis, a detailed
description of the instrument configuration is necessary. Figure 6 is an exploded view of the thermometer showing its laminated construction and internal radiation shield; when bonded
together, the sandwich is 0.9 mm thick. The DTREM package,
shown pictorially in Fig. 1 and in exploded view in Fig. 2,
is a cube of G-10 fiberglass, 36x32x41 mm, with walls 2.3 mm
thick. The top of the package is covered by a kovar sheet
to which three 10x20-mm solar cells are attached for the dust
and radiation measurements, conducted by S. Freden of the
Manned Spacecraft Center and B. O'Brien of the University of
Sydney.[4] This kovar sheet has two thermistors attached under-
neath, which are on scale only above 305°K. The solar-cell
outputs are used in the thermal analysis to give the isolation
directly, since parts of the EASEP spacecraft shadow the DTREM
package, as seen in Fig. 4. The north face of the DTREM package is covered by the nickel thermometer but separated from it
by 20 layers of superinsulation (about 2 mm thick), which are
pierced by a pair of thermal isolator standoffs and nylon
bolts. The DTREM is held by 2 bolts to an insulating fixture
attached to the EASEP primary structure.

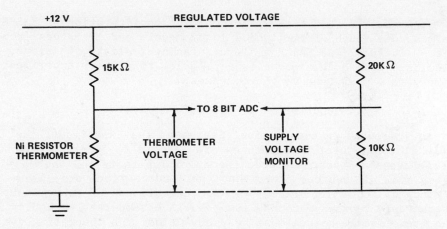

Fig. 5 Thermometer schematic circuit diagram (all precision
        resistances <25 ppm/°K).

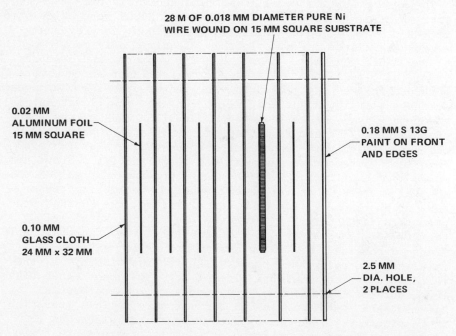

Fig. 6 Fiberglass nickel resistance thermometer, exploded
        view (epoxy binder not shown).

Figures 3, 7, and 8 show the deployed configuration of EASEP, almost due south of the landed lunar module and 18 meters from it.[5] The deployed EASEP is level, facing a slight, 4° slope, inclined upward toward the lunar module. The laser experiment array is 4 m distant to the northwest.

Description of the Data Analysis

As indicated in the section, Sketch of the Experiment, the nickel thermometer temperatures are reduced to lunar surface brightness temperatures once the power inputs to the thermometer from all sources other than the lunar surface are calculated and added algebraically. The calculation of these inputs requires a knowledge of the lunar surface temperature, and so an iterative technique must be used. Also, the insulating qualities of the DTREM package and of the thermometer laminates require a multinode analysis to deal with the large thermal gradients. The high lunar vacuum insures that the heat transfer is dominated by thermal radiation.

The method used is as follows. The geometric view factors from the thermometer to various parts of EASEP were calculated by the standard computer program CONFAC[6] and, as previously noted, are listed in Table 1; these required only knowledge of the EASEP geometry. From the EASEP geometry and orientation, the dimensions of its shadow on the lunar surface were obtained; hence, the geometric view factor from the thermometer to the sunlit lunar surface, F_{NiS}, for each sun elevation, as given in Fig. 9, was calculated. The DTREM package was then divided into 270 thermal modes for steady-state analysis, and the thermal analysis computer preprocessor Chrysler Improved Numerical Differential Analyser (CINDA[7]), and its subroutines, were used to write a FORTRAN computer program that solves the thermal network, that is, calculates the equilibrium temperature of every node. For each sun elevation, the FORTRAN program is given 8 EASEP temperatures, including the nickel thermometer temperature, and the geometric view factors. A lunar surface temperature is assumed, and each side of Eq. (1) is calculated using the iterative subroutine CINDSM. If the calculated values on either side of Eq. (1) do not agree within acceptable limits, a new lunar surface temperature is assumed, and the CINDSM analysis is repeated. This iteration is continued until an acceptable energy rate balance is obtained for the nickel thermometer with its temperature fixed at the measured value. As indicated, the foregoing iteration is repeated for selected values of sun elevation angle.

The EASEP temperatures used as input to the analysis were from thermometers on the east, west, and bottom of the EASEP

Fig. 7 EASEP, deployed configuration, showing the lunar
        module (18 m due north), the laser experiment array,
        and the "rock." Note the lunar surface north of
        EASEP between the solar panels.

Fig. 8   EASEP, deployed configuration, showing solar panels,
         antenna, seismometer, and "rock."   The nickel
         thermometer of DTREM I views the lunar surface
         between the solar panels out to the "rock."

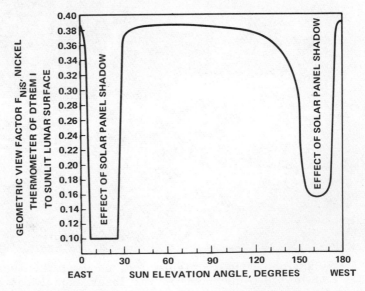

Fig. 9   Geometric view factor F_{NiS}, for the thermal radiation
         calculation.

primary structure (the 1.2-mm-thick aluminum box in contact with the lunar surface), on the east and west solar panels, on the northeast and northwest of the central station thermal plate, on the DTREM kovar sheet (when on scale), and the DTREM nickel thermometer (see Figs. 2 and 3). The temperature of the dark lunar surface was taken at 200°K. Since a value of 0.076 for the lunar surface reflectance A was measured on an Apollo 11 soil sample,[8] the initial value of the (iterated) lunar surface temperature was taken to be $[(1-A) S \sin\phi /\sigma]^{\frac{1}{4}}$, where S is the solar constant and ϕ the sun elevation.

The heat rate balance for the nickel thermometer coil of wire then included radiation from the black sky, the sunlit lunar surface, the dark lunar surface, the west side of the bracket, the back of the bracket, the boom, astronaut handle and carry handle, the east solar panel, the west solar panel, and both parts of the rock (this is the split rock so evident in Figs. 7 and 8); a conduction contribution from 8 nodes on the laminated part of the thermometer and Joule heating; and a direct solar input to the thermometer due to the tilt of the DTREM and sunlight reflection from the lunar surface, the bracket, and the back of the solar panels. The calculation of internode conductances and of the heat rate input to each node was not unusual; the constants used in the analysis are given in Table 2. Of interest may be the fact that calibrated solar cell outputs were available as measured insolation of the DTREM package so that calculation of EASEP shadow dimensions on the DTREM structure was unnecessary. The algebraic sum of these heat rates should be zero, and iteration was continued until the sum was less that 1/2% of the lunar surface input. These heat rates, grouped into nine independent partial sums, divided by the heat rate from the sunlit lunar surface, are plotted in Fig. 10.

Referring to Fig. 10, we see that, aside from the black sky, the largest correction to the data is from the EASEP bracket, typically 30% but rising to 120% at low sun elevation. In estimating the lunar surface temperature error, we _assume_ that the calculated corrections of Fig. 10 are independent and accurate to a 10%, relative 1 sigma error. Then, since the relative temperature error is 1/4 of the relative error in the radiated energy, we estimate the temperature error (1 sigma) to be +4°K at 90° sun elevation and ± 20°K at 20° and 160° sun elevation.

The calculated lunar surface brightness temperature is given in Fig. 11, with the estimated error indicated. A curve of $[(1-A) S \sin\phi /\sigma]^{\frac{1}{4}}$, the Lambert temperature for A=0.076, is

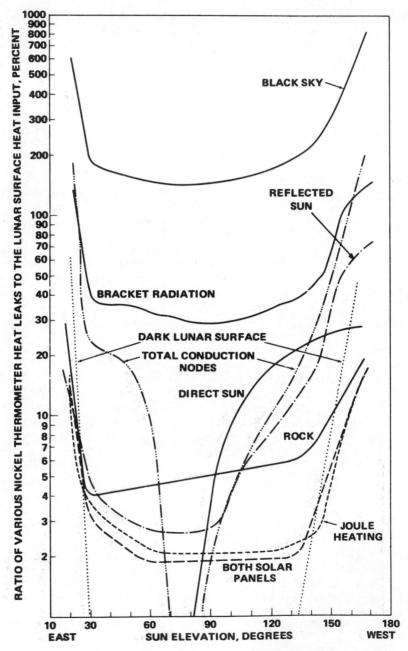

Fig. 10 Calculated heat leaks used in the data reduction
         [Eq. (1)]. The lunar surface heat input is
         $\varepsilon_{Ni} F_{NiS} \varepsilon_S \sigma T^4 S$.

Table 2 Constants used in thermal analysis[a]

	Absorptivity α	Emissivity ϵ	Conductivity k,mw/m-°K
S-13G thermal paint	0.288	0.885	350
Solar Cell	0.81	0.83	...
Kovar (painted)	0.90	0.90	...
Solar Panel (back)	0.3	0.3	...
Rock	1.0	1.0	...
Dust	1.0	1.0	...
G-10 fiberglass, nylon	...	...	350
Vespel	...	...	380
Aluminum	...	...	116,800
Kovar	...	...	29,700
Nickel	...	...	62,300

[a]Solar Constant, S=1350 w/m^2; Stefan-Boltzmann constant $\sigma= 5.67 \times 10^{-8}$ w/m^2 - °K^4; reflectance of lunar surface A=0.076; sun declination, deg., 1.3 - 0.022 (day-200); Apollo 11 longitude, latitude, 23° .48, + 0°.663.

also given in Fig. 11 for comparison. ("The Lambert temperature is the temperature a perfectly insulating Lambert surface with unit emissivity would come to if it absorbed the same amount of radiation as the surface under consideration."[1])

As an illustration of the calculated temperatures, we note that at a sun elevation of 20° the sun is at 70° incidence to the east face of the DTREM package, whereas the solar cell thermistors on top are not yet on scale and the west face is, of course, completely in shadow. At this sun elevation, the nickel temperature is 273°K, whereas other nodes on the thermometer range from 10°K above this to 0.3°K below, the north face of the DTREM package is typically 5°K hotter, the east face 19°K hotter (the range is 10 to 60°K), the top solar cells 45°K hotter, and the west face about 20°K colder. These numbers illustrate the advantage of an additional thermometer to measure the conductive leak from the nickel thermometer.

Results of the Measurement

The nickel thermometer data from Apollo 11 are given in Fig. 4, and the lunar surface brightness temperatures calculated from these data are given in Fig. 11 with the estimated 1 sigma errors. The Lambert temperatures for a reflectance of 0.076, given in Fig. 11, are seen to represent the brightness temperatures adequately. Table 3 lists Lambert temperatures for selected values of the sun elevation as well as theoretical

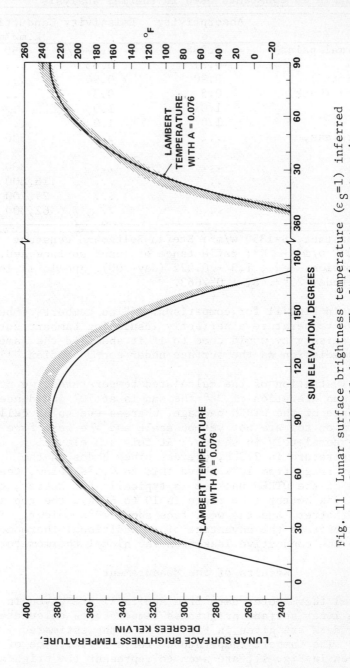

Fig. 11   Lunar surface brightness temperature ($\varepsilon_S=1$) inferred
from Apollo 11 data.   The Lambert temperature is
$[(1-A) \ S \ \sin\phi/\sigma]^{\frac{1}{4}}$.

Table 3 Theoretical lunar surface brightness temperatures, °K, for selected values of the sun ele-
vation angle and the thermal parameter $\gamma = (k\rho c)^{\frac{1}{2}}$,
$cm^2\text{-}sec^{\frac{1}{2}}\text{-}°K/g\text{-}cal$

Sun elevation, deg.	Thermal Parameter γ				
	250	500	750	1500	∞[a]
10	228	236	240	244	244
20	284	289	291	292	289
30	317	320	321	322	318
90	383	384	384	375	378
150	324	324	324	324	318
160	297	295	295	295	289
170	257	253	252	250	244
180	183	161	149	130	...
190	165	143	131	112	...
270	133	114	104	88	...
360	122	104	94	80	...

[a]This limiting case is the Lambert temperature for A=0.076

values of lunar surface temperatures calculated with the homo-
geneous-moon model of Jaeger[9] for various values of thermal
parameter $\gamma = (k\rho c)^{-\frac{1}{2}}$. Comparison of Fig. 11 and Table 3
suggests that the results are adequately represented by the
homogeneous model for any value of γ greater than $500cm^2\text{-}sec^{\frac{1}{2}}\text{-}$
°K/g-cal. The Surveyor results are usually higher than this
value, in the range of 600 to 1000. Table 3 includes also a
number of predicted temperatures for angles after sundown
(assuming no temperature dependence of thermal conductivity
k or specific heat capacity ρc), to point out the relative
insensitivity of the daytime data to γ and the distinct advantage of lunar night data for discriminating between γ values.

Figure 9 shows depressions in the thermometer-to-sunlit surface geometric view factor at low sun angles. These depressions are the result of shadows of the solar panels on the
lunar surface. Note that, of the total value of 0.38 for the
geometric view factor, fully 74% is attributable to the lunar
surface between EASEP and the end of the solar panels, an area
of about 1 m^2, which is clearly visible in Fig. 8, and to a
lesser degree in Fig. 7. This surface is seen to be reasonably flat and free of boulders and depressions, although it
has a number of foot imprints.

As mentioned in the section, Sketch of the Experiment, comparison of nickel temperatures from successive lunations was expected to bear on the degradation of thermal finishes used or

viewed, degradation due to ultraviolet damage to the paint, and/
or dust accretion on the paint. We observe that the available
data of the second lunation reproduce that of the first day
within the digitizing error of 1.6 K which corresponds to a 1.6%
change in the lunar surface input to DTREM at lunar noon or,
equivalently, corresponds to a 10% change in absorptivity or an
absolute change of 0.03, possibly because of the addition of 3%
dust. A dust accretion rate lower than 3%/month is not surprising. Also, degradation of the thermometer surface by ultraviolet light is not expected at the low irradiation exposure of
north-facing surfaces.

The preflight thermal data, used in the analysis, were not
affected by adding up to 10% dust to the thermometer surface;
i.e., our results are also consistent with negligible dust
accumulation on the thermometer during lunar module ascent.
To quantify this result, we assume that, if the inferred lunar
temperatures were all high, which implies at least a 1 σ
increase, we should attribute this increase to changes in our
preflight thermal parameters. The 1 σ or 4°K increase would
correspond to 7% dust cover. These results imply that the
nickel thermometer does not have good resolution for dust
detection, either for absolute determination or for detection
of short-term changes.

References

[1]Shorthill, R. W., "Infrared Moon: A Review," Journal of
Spacecraft and Rockets, Vol 7, No. 4, April 1970, pp. 385-397.

[2]Stimpson, L. D., Hagemeyer, W. A., Lucas, J. W., Popinski, Z.,
and Saari, J. M., "Revised Lunar Surface Temperatures and Thermal Characteristics from Surveyor," Analysis of Surveyor Data,
TR 32-1443, June 30, 1969, Jet Propulsion Laboratory.

[3]Montgomery, C. G., Saari, J. M., Shorthill, R. W., and
Six, N. F., "Directional Characteristics of Lunar Thermal
Emission," TN R-213, Nov. 1966, Research Laboratories, Brown
Engineering Company, Inc., Huntsville, Alabama.

[4]Bates, J. R., Freden, S. C., and O'Brien, B. J., "The Modified Dust Detector in the Early Apollo Scientific Experiments
Package," The Apollo 11 Preliminary Science Report, SP-214,
p. 199, 1969, NASA.

[5]Shoemaker, E. M., Bailey, N. G., Batson, R. M., Dahlem, D.H.,
Foss, T. H., Grolier, M. J., Goddard, E. N., Hait, N. H.,
Holt, H. E., Larson, K. B., Rennilson, J. J., Schaber, G. G.,

Schleicher, D. L., Schmitt, H. H., Sutton, R. L., Swann, G. A., Waters, A. C., and West, M. N., "Geologic Setting of the Lunar Samples Returned by the Apollo 11 Mission," The Apollo 11 Preliminary Science Report, SP-214, P. 41, 1969, NASA.

[6]Toups, K. A., "A General Computer Program for the Determination of Radiant-Interchange Configuration and Form Factors - CONFAC II, TR SID 65 1043-2, Oct. 1965, North American Aviation, Inc., Space and Information System Division.

[7]Lewis, D. R., Gaski, J. D., and Thompson, L. R., "Chrysler Improved Numerical Differencing Analyzer for 3rd Generation Computers," TN-AP-67-287, Oct. 20, 1967, Chrysler Corporation Space Division, New Orleans, La.

[8]Birkebak, R. C., Cremers, C. J., and Dawson, J. P., "Thermal Radiation Properties and Thermal Conductivity of Lunar Material," Science, Vol. 167, No. 3918, 1970, pp. 724-726.

[9]Jaeger, J. C., "Conduction of Heat in a Solid With Periodic Boundary Conditions, with an Application to the Surface Temperature of the Moon," Proceedings of the Cambridge Philosophical Society, Vol. 49, Part 2, April 1953, pp. 355-359.

DEVELOPMENT OF AN IN SITU THERMAL CONDUCTIVITY MEASUREMENT FOR THE LUNAR HEAT FLOW EXPERIMENT

Marcus G. Langseth, Jr.[*]
Lamont-Doherty Geological Observatory of Columbia University
Palisades, New York

Elisabeth M. Drake[†] and Daniel Nathanson[‡]
Arthur D. Little, Inc., Cambridge, Massachusetts

James A. Fountain[§]
NASA Marshall Space Flight Center, Alabama

List of Symbols and Units

$\vec{F}$ = heat flow vector, W/cm^2
k = thermal conductivity, $W/cm-K$
T = absolute temperature, K
z = vertical distance (positive downward), cm
To = initial temperature, K
Q = power generated, W
a = radius of spherical source, cm
α = thermal diffusivity, cm^2/sec
$erfc(x)$ = the complementary error function $2\pi^{-1/2} \int_{x}^{\infty} e^{-u^2} du$

γ = $(k\rho c)^{-1/2}$, thermal parameter, $cm^2 K/W-sec^{1/2}$
σ = standard deviation about theoretical curve
p = density, gm/cm^3
c = heat capacity, $W-sec/gm-K$

The work described in this paper was performed under NASA Contracts NAS 9-6037 with Columbia University, and NAS 9-5829 with Bendix Aerospace Systems Division of the Bendix Corporation, and subcontract SC 0242 with Arthur D. Little, Inc.. A. Wechsler, F. Ruccia and R. Merriam made important contributions to the experimental and theoretical programs at ADL. The measurements at NASA Marshall Space Flight Center were conducted with the help of E. West and R. Scott. The fundamental concepts of the conductivity experiment are due in large part to S. P. Clark, Jr. of Yale University. L-DGO Contrib. No. 1678.

[*]Senior Research Associate.
[†]Project Staff.
[‡]Staff Associate.
[§]Physicist.

Introduction

On Apollo Missions 15, 16 and 17 temperature and thermal conductivity measurements will be made in the lunar surface layer to determine the steady-state heat flow through the surface of the moon. If local, near surface effects are not large, the net surface heat flux represents the heat loss from the deep interior. For the moon, the heat loss is expected to be principally dependent on the total abundance of heat generating long lived radioisotopes. These isotopic abundances in turn set important constraints on the bulk composition of the moon. Thus the principal objective of the heat flow measurements is to gain evidence on the bulk chemistry of the moon. The heat flow experiment, described in more detail by Langseth et al.[4], is part of the Apollo Lunar Surface Experiments Package (ALSEP).

The heat flow is determined by making independent measurements of the vertical temperature gradient, $\partial T/\partial z$, and the thermal conductivity, k, in the lunar surface layer, the regolith. If the heat flow is purely conductive, its value is given by the product of these two measurements: $\vec{F} = k \, \partial T/\partial z$. In the lunar regolith the assumption of conductive heat flow is not strictly true. Interparticle heat transfer by radiation is important. However, if we assume an effective conductivity that includes conductive and radiative transfer, we can use the steady state conductive theory with only very small errors if the gradients are small.

It is clear from the expression that errors in measuring either parameter affect the accuracy of the heat flow determination an equal amount. In this paper we will discuss the methods of determining thermal conductivity in the lunar surface layer, and the errors associated with the experimental technique.

The determination of thermal conductivity has always been the largest source of error in measuring heat flow from the interior of the earth. The thermal properties of the layers of rock or sediment over which the gradient is measured are usually heterogeneous and present the investigator with difficult sampling problems. Secondly, both laboratory and in situ techniques of conductivity measurement present experimental difficulties that can contribute substantial errors. The measurement of conductivity of the lunar regolith presents similar problems; in addition the limited weight and time available on the lunar missions do not allow the most ideal sampling procedures or experimental design.

The highly pulverized nature of the regolith and the vacuum condition at the moon's surface should produce an extremely low thermal conductivity for the bulk of the near surface material, similar to that of evacuated powders tested on earth (see Wechsler et al.[9]). Tests of Apollo 11 fine grained samples from the regolith by Birkebak et al.[1] indicate a conductivity of 1.71×10^{-5} W/cm-K at 205 K (see chapter 3.b.). In contrast, the thermal conductivity of solid rock fragments in the regolith is 1.63×10^{-2} W cm^{-1}-K^{-1} based on the work of Robie et al.[6] and Horai et al.[3] The ratio of the two conductivities is very nearly three orders of magnitude!

The low thermal conductivity of the regolith layer presents some advantages for a heat flow measurement. Even very low values of heat flow induce relatively large gradients in the regolith that can be easily detected. A second advantage is that the extremely low thermal diffusivity causes rapid attenuation with depth of the large surface temperature variations during a lunation. The lunation temperature cycle with a surface amplitude of $\sim$300 K is virtually undetectable at a depth of one meter based on the conductivity values of regolith fines above and is acceptably small for k values up to 100 times larger.

On the other hand, very low conductivity presents some serious difficulties in the heat flow measurement. First, any instrument emplaced in the soil will distort the existing subsurface thermal regime it is designed to measure. The probes we use in the Apollo measurements are designed to have as low a conductance as practicable; even so, they will shunt the heat flow in the subsurface to some degree. Furthermore, because of the low thermal diffusivity of the surface layer, extremely long times are required for the probe and regolith to come to a new equilibrium condition after emplacement. However, these effects can be rather easily corrected for using conduction theory provided we know the average bulk thermal conductivity of the surrounding material.

The Heat Flow Experiment Design

The problems suggested above due to the heterogeneous regolith are made tractable by redundancy in the experiment design: two gradient measurements and four separate k measurements are made in each of two separate holes 5 m apart. The vertical temrature gradient and thermal conductivity measurements will be made in the regolith zone 2 to 3 m below the surface. Two 3 m sections of hollow drill rod called borestem approximately 2.5 cm in diameter will be drilled into the regolith by an

astronaut.   The borestem, which is made of thin-walled epoxy
fiber glass tubing reinforced with axially aligned boron, will
be left in the lunar soil after drilling is complete, and two
identical heat flow probes will be placed at the bottom of each
of the two boreholes.   The configuration of one probe in the
regolith is shown in Fig. 1.

The identical heat flow probes are emplaced in the regolith
about 10 m apart and are connected by cables approximately 8 m
long to an electronics box on the surface that provides switch-
ing, signal conditioning, and programming for the probes.   The
experiment is designed to transmit data from the moon for a
period of two years after emplacement.

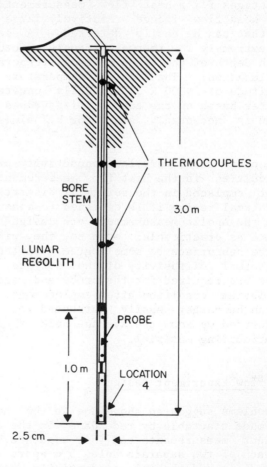

Fig. 1 Configuration of a heat flow probe emplaced in the
lunar subsurface.

Each heat flow probe consists of two almost identical 50 cm long sections that are connected by electrical cables.   Each section contains a gradient sensor bridge, a conductivity or ring sensor bridge, and two heaters. The gradient bridges provide the primary temperature difference measurements for determining the vertical thermal gradient.   The ring bridges are used in conductivity measurements and provide additional thermal gradient data. The four arms of these bridges are platinum resistance elements.   Adjacent arms of the bridge are located in sensor cans at opposite ends of the 50-cm probe section; consequently, when a voltage is applied across the bridge, the output is a measure of temperature difference over the section. Bridge current combined with the applied voltage gives a measure of the average absolute temperature of the sensor elements.

There are also four chromel-constantan thermocouple junctions in the cable located at distances of 0. 65, 115 and 165 cm from the top of the probe. These junctions measure the absolute temperature in the upper two meters of the borehole. Measuring ranges and accuracies for the heat flow probe measurements are shown in Table 1.

Table 1   Range and accuracies of heat flow
experiment measurements

	Probe Temp., K (lower meter)	Lunar subsurface Temp., K (upper 2 meters)	Temperature difference, K in lower meter for 50-cm length	Thermal Conductivity W/cm-K
Range	200-250	90-350	± 20	2×10^{-5} to 4×10^{-3}
Resolution	0.1	0.5	0.001	$\pm 20\%$
Accuracy	± 0.1	± 0.5	± 0.003	$\pm 20\%$

The primary method to determine thermal conductivity will
be in situ experiments in the interval 2-3 m below the surface.
In principle the measurement is made by monitoring the response
of either the gradient or ring sensor to initiation of a heater
surrounding the gradient sensor.   The heater can be turned on
and off by commands transmitted from Earth.   Each probe section
contains two heaters so that there is the potential of measuring the conductivity at eight different locations in the
regolith.

Accurate measurement of thermal conductivity of evacuated
powders presents great difficulties in earth laboratories.   The
measurements in the lunar regolith present further difficulties
mainly because the geometry of the probes necessary to assure
emplacement of the experiment in the lunar surface is not optimum for conductivity measurement.   The complex geometry of the
probes in borestems required that the theory of the experiment
be developed with detailed finite-difference models.   These
models are our prime means of reducing the data.

An extensive experimental program has been carried out to
test the design.   Conductivity measurements have been made under
simulated lunar thermal conditions in materials with conductivities in the range expected on the moon.   These experiments have
also been used to validate our theoretical models.

The errors in these measurements are inherently large.   At
the start of the development program we set a rather realistic
goal of ±20% accuracy for the conductivity measurements.   This
would allow a heat flow determination accuracy within about ±25%
(the estimated maximum error of gradient measurement is ±5%).
The actual error in the lunar measurement will depend on many
as yet unknown factors (e.g., the heterogeneity of regolith,
the mechanical disruption of the regolith during drilling and
the actual thermal conductivity).   We have therefore taken more
than the usual amount of care in fabricating and testing those
parts of the experiment that we can control.

If the experiment operates as planned, two independent determinations of conductivity will be made at each of eight locations in the regolith in two widely separated groups of four.
To these results we can add laboratory measurements on returned
samples from a nearby borehole and observations of thermal waves
propagating downward from the surface due to the annual and
monthly variation in solar flux. The main function of the thermocouples in the cable is to detect such thermal waves.

In this paper we describe in detail the design of the conductivity experiment and a simplified analytical model is presented

to show the approximate relationships between sensor response
and thermal conductivity. This discussion leads into a presentation of the detailed finite-difference model and some typical
results. The experimental program is described next and comparisons between experiment and theory are presented. Lastly, we
discuss the measurement procedure and interpretation technique
we plan to use when the experiment is on the moon.

Design of the Conductivity Experiments

Low Range of Thermal Conductivity (Mode 2)

Figure 2 shows the end of a heat-flow probe section as it
might appear in the lunar subsurface. To make measurements in
the lower range of anticipated regolith conductivities (from
1.5×10^{-5} to 5×10^{-4} W/cm-K), the heater is activated at
0.002 watt, and the temperature rise of the underlying gradient
bridge sensor is observed. This experiment is related to point
source techniques of conductivity measurement; however, thermal
links between the heater and sensor, as well as the heater and
the surrounding lunar material, are complicated by constructional details and emplacement geometry (see Fig. 2). Thermal
couplings between the heater and sensor are by radiation between the outer sheath and sensor can and by conduction through
mounting bushings at either end of the can. Heat flow to the
lunar material is across a radiation gap to the borestem. In
addition, appreciable heat flows axially along the probe and
borestem walls. Hence the temperature rise at the sensor depends in a complex way on the thermal linkages in the probe and
the in situ geometry.

High Range of Thermal Conductivity (Mode 3)

For the high end of the anticipated lunar conductivity
range (from 3×10^{-4} to 5×10^{-3} W/cm-K), the same heater wind-
ing is energized at 0.5 W, and temperature response is measured
at the ring bridge sensor about 10 cm away (see Fig. 2). The
ring bridge sensor, which has a much lower thermal capacity
than the gradient sensor, is thermally coupled to the heater by
two paths: directly through the thin-walled probe sheath and
radiatively through the borestem and lunar material. The borestem and probe components have short thermal time constants relative to the lunar surroundings over the design measuring range.
Consequently, long-term response of the ring sensor, >3 hours,
gives a measure of the thermal properties of the lunar subsurface.

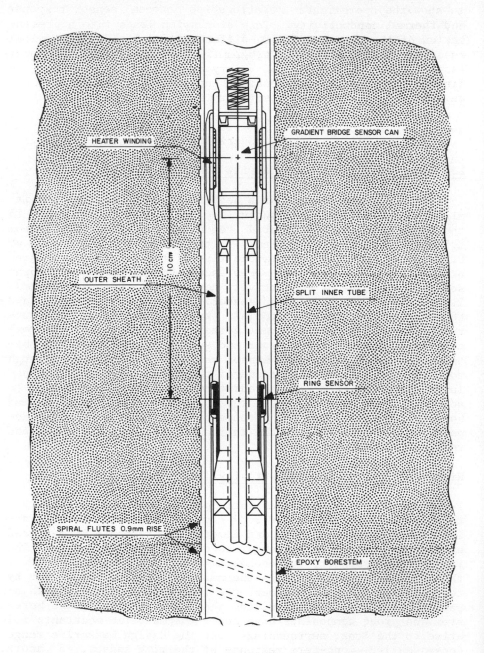

Fig. 2 Cutaway view of the end of a probe section in the boron reinforced epoxy borestem as it might appear in the lunar subsurface.

Mathematical Modeling of the Experiment

A Simplified Analytical Model

It is instructive to examine briefly a simplified model
of the experiment to see qualitatively how the temperature rise
is related to the thermal properties of the surrounding lunar
medium. The model we choose is a continuous spherical surface
source of radius, a, in an infinite, homogeneous and isotropic
medium. Heat is liberated on the surface at a constant rate Q,
at times greater than zero. Carslaw and Jaeger[2] give the temperature distribution outside such a source as a function of
time t and radial distance r. It can be written

$$T(r,t)-T_0 = \frac{Q}{8\pi k r a}\left\{\frac{2(\alpha t)^{1/2}}{\sqrt{\pi}}\left[\exp\left(\frac{-(r-a)^2}{4\alpha t}\right)-\exp\left(\frac{-(r+a)^2}{4\alpha t}\right)\right.\right.$$
$$\left.\left.-(r-a)\,\mathrm{erfc}\,\frac{(r-a)}{2\sqrt{\alpha}t}+(r+a)\,\mathrm{erfc}\,\frac{(r+a)}{2\sqrt{\alpha}t}\right\}\right.$$

The rate of temperature rise can be written

$$\frac{\partial T(r,t)}{\partial t} = \frac{Q\,Y}{8\pi^{3/2}r a\, t^{1/2}}\left[\exp\left(\frac{-(r-a)^2}{4\alpha t}\right)-\exp\left(\frac{-(r+a)^2}{4\alpha t}\right)\right]$$

These models do approximate the actual experiment, since
laboratory tests have shown that temperature difference over the
radiation gap between the heater and the borestem walls become
nearly constant about 4 hrs after heater turn-on, and subsequently heat is being supplied at a nearly constant rate to a small
band on the perimeter of the borestem. In our qualitative model
we replace the band on the borestem with the spherical surface
source.

The above expressions are more illustrative if we expand
the exponential and complementary error function in powers of
the quotients $(r - a)/2\sqrt{\alpha t}$, and $(r + a)/2\sqrt{\alpha t}$. The solutions
become

$$T(r,t)-T_0 = \frac{Q}{4\pi k}\left(\frac{1}{r}-\frac{1}{\sqrt{\pi\alpha t}}+\frac{(r^2+a^2)}{12\sqrt{\pi}(\alpha t)^{3/2}}-\cdots\right) \qquad (1)$$

and

$$\frac{\partial T(r,t)}{\partial t} = \frac{Q\gamma}{16\pi^{3/2}\alpha} \left(\frac{1}{t^{3/2}} - \frac{(r^2+a^2)}{4\alpha\,t^{5/2}} + \cdots \right) \qquad (2)$$

To model the Mode 2 experiment we put $r = a$ in which case the third term of Eq. 1 becomes small more rapidly than the second term for increasing t. It is clear that $T - T_o$ approaches $Q/4\pi ka$ as t approaches infinity; the steady-state solution for a spherical source. At steady-state the temperature rise is inversely proportional to conductivity. However, for low thermal conductivities, i.e., for $k < 4 \times 10^{-4}$ W/cm-K, the second term diminishes very slowly with time, becoming small when, $\alpha t \gg a^2$. For a conductivity of 4×10^{-5} W/cm-K more than 3×10^5 seconds are required before the first term becomes dominant. Thus we expect the Mode 2 experiment in lunar regolith material to require tens of hours to perform and even then the sensor will not reach steady state. It is possible to estimate the steady-state value from the transient response at long times from Eq. 2 when the slope is approximately proportional to $t^{-3/2}$.** For $\alpha t \gg a^2$ the asymtotic value of $T - T_o$ can be estimated from observations of slope up to time t' by

$$(T-T_o)\Big|_{t\to\infty} = (T-T_o)\Big|_{t=t'} + \int_{t'}^{\infty} \frac{\partial T}{\partial t}\,dt$$

However, this extrapolation is very vulnerable to errors due to transients in the surrounding soil unrelated to the heater turn-on, and errors in measuring $\partial T/\partial t$ at long times. These rates of temperature rise are typically a few millidegrees per hour.

A second technique based on Eq. 1, which is more reliable, is to take advantage of the fact that as time increases, $T-T_o$ becomes increasingly less sensitive to ρc. Consequently, if with the aid of a more precise model and a specific ρc, we determine k for increasingly longer times, it will approach an asymptotic value equal to the true k.

**We will show later that based on experimental data and a more exact finite-difference model of the experiment that at long times the slope is well approximated by an expression of the form $\partial T/\partial t = Ct^\beta$. The exponent, β, is a function of conductivity of the medium. The value of β for $k = 4 \times 10^{-5}$ W/cm-K is about -1.3.

Lastly, the slope can be interpreted in terms of the thermal parameter, γ, for long times when the second term in Eq. 2 is small. The thermal mass ρc can usually be estimated to $\pm 25\%$, thus allowing an estimate of k alone.

Note that the spherical model describes the temperature at the perimeter of the borestem adjacent to the heater. The heater and gradient sensor are thermally separated from the borestem wall by a radiation gap. For values of k above 10^{-4} W/cm-K, the temperature rise of the gradient sensor is principally determined by the radiative gap and the experiment becomes less sensitive to the thermal properties of the surrounding material. For values of $k > 4 \times 10^{-4}$ the uncertainties in interpretation of the Mode 2 experiment become too large for reliable use, and the Mode 3 technique must be used. The radiative transfer between the heater and the borestem walls results in a strong interdependence between temperature rise of the sensor and absolute temperature.

The simple spherical surface model is a very poor model of the Mode 3 experiment because of the appreciable flow of heat along the probe. Nonetheless we can see that when $r \cong 10a$, the terms in Eqs. 1 and 2 containing $r^2 + a^2$ become much more important. Since temperature rise at the ring sensor does not approach steady-state for practical times after heater turn-on, we use transient data on the temperature rise and slope and more precise mathematical models of the experiment to determine the value of $[k\rho c]^{-1/2}$. Both the temperature rise and slope are nearly unique functions of this parameter. To determine k we depend on independent estimates of density and specific heat which, as noted above, should give k to $\pm 25\%$.

A More Precise Model; The Finite-Difference Approach

Next we will describe a far more exact model of the Mode 2 and Mode 3 experiments using finite-difference methods. The relations between temperature rise and the thermal parameters of the probe and the lunar medium are computed by numerical techniques. These models are of such complexity that they closely describe the actual physical experiment. Consequently, they are used to develop empirical relations between temperature rise and k, ρ and c of the lunar medium.

A simulation of the thermal behavior of a physical system, using finite-difference techniques, generally requires the following steps: 1) geometrically subdividing the physical system into thermal zones or nodes; 2) describing the thermal parameters for each zone and defining the heat-balance equations for the zones or nodes; 3) solving the equations for node temperatures. Accurate thermal modeling and proper selection of subdivision size are of particular importance in the thermal simulation of the lunar heat flow probe and lunar material because a small error in computed temperature could result in a substantially large error in the inferred thermal conductivity. For example, in the Mode 2 experiment the conductivity of the surrounding medium is related to a small temperature rise at the gradient sensor -- between 0.3 and 1.0 K.   A 10% change in the sensor temperature rise corresponds to a 50% change in thermal conductivity, for values in the range of 2×10^{-5} to 4×10^{-4} W/cm-K.   Therefore, prior to the development of complete models for the heat flow probe and surrounding lunar medium, preliminary studies were made for the following purposes: 1) to determine the size of the finite-difference subdivision in the low-conductance probe sheath necessary to render an accurate description of the heat flow between sensor locations; and 2) develop finite-difference models and procedures for describing an infinite lunar medium.

The heat flow equations for the thermal models were written in finite-difference form using the "zone method" of Strong and Emslie.[7]   In the "zone method" the temperature within a zone is approximated to second order in terms of the space coordinates.   Based on the second order assumption, approximate formulas are derived expressing the heat flow and boundary conditions in terms of the mean temperature of the zone and the boundary temperatures.

The heat-balance equations for each zone were solved on a CDC 6600 digital computer, using an existing thermal analysis computer program designed to solve systems of nonlinear differential equations.   At each time step, the simultaneous heat-balance equations are solved for the temperature distribution by a modified Gauss-Seidel procedure using the Newton-Raphson iteration method. Because of the iterative procedure, the individual equations are solved to the machine's tolerance and the system of equations is solved to a prescribed tolerance which is defined as the sum of the absolute value of the residuals of each equation. The computer program uses an implicit procedure for the integration in time.

Computer models were developed to describe the thermal performance of the probes during the two modes of conductivity measurement, accounting for specific thermal characteristics (i.e., cabling, probe stop at the bottom of the borestem, etc.) at each of the four experiment locations.   The details of a thermal model subdivision for a typical experiment location are illustrated in Fig. 3a. Many of the construction details in

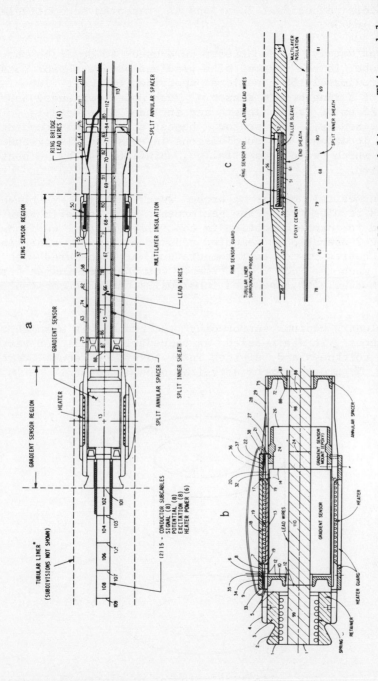

Fig. 3 a) Subdivision of a probe section end for finite-difference modeling. This model corresponds to location 1. The numbers identify zones and boundaries. b) Detail of gradient sensor finite-difference model. c) Detail of finite-difference model in the vicinity of the ring sensor.

the region of the gradient and ring bridge sensors are similar at all four experiment locations as presented schematically in Fig. 3b and 3c.

Approximately 40 equations were used to describe the heat flow probe in the region of a gradient bridge sensor location and approximately 50 equations described the heat flow in the region of a ring bridge sensor. Each model extended approximately 25 cm of probe length and included the details of the gradient sensor, ring sensor, heater and thermal couplings at a given location. The subdivisions in these models extended to regions which were essentially uninfluenced by heater activation.

The numerical models of probe were verified by comparing computed results from probe performance with experimental data obtained for probe operation in controlled thermal environments. For Mode 2 operation, computed results of a probe radiating to a black cavity at constant temperature were compared with test results when the Mode 2 experiment was operated inside a thick aluminum tube with blackened inner walls, held at constant temperature.

The lunar medium surrounding the probe was assumed to be homogeneous. It was modeled as concentric annuli of varying radial thickness and divided into layers of equal vertical height. Figure 4 depicts a typical annular element.

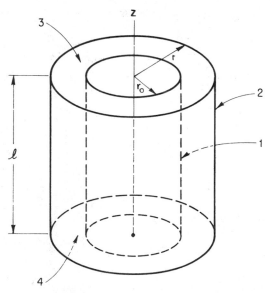

Fig. 4   Annular cylindrical element of finite-difference model of lunar medium.

Expressions for the net heat flow from faces 1 through 4 are found in terms of the average surface temperatures, T_1 through T_4, and the mean temperature of the element, T_m. Thermal connections between the zones are made by matching the net heat flows at zone boundaries. For example the heat flow from the outer surface of an element (surface 2 in Fig. 4) is equal to the negative of the heat flow from the inside surface of the surrounding annular ring.

The vertical height of each annular element was chosen to be a constant whose value depends on the thermal properties of the medium and the operating mode of the conductivity experiment. Studies showed that the accuracy of calculations could be enhanced if two separate models of the lunar medium were developed--one tailored for simulation of the Mode 2 experiment (lower power, lower conductivity) and the other for the Mode 3 experiment (higher power, higher conductivity).

Because of the rapid decrease of thermal gradients with distance from the borestem, we can increase the thickness of each zone with the radial distance r without decreasing the precision. The radial temperature distribution in a medium for a spherical-surface heat source varies approximately at $1/r$ (see Eq. 1). Hence, in the finite-difference models the outer radius of the nth annular element is given by $r_n = C^n r_0$, where C is a constant and r_0 is the inner radius of the borestem.

The number of vertical subdivisions, the number of concentric annuli, and values of the quantities C and ℓ defined the geometrical configuration of the medium and the number of locations or nodes that required descriptive heat flow equations. The values of these parameters which were used to define the surrounding medium are summarized in Table 2.

At the outer boundary of the lunar medium the heat flows from the surfaces are calculated from the analytical solution for a continuous spherical surface heat source given earlier.

Prior to incorporating the finite-difference representation in the lunar medium with that for the probe to form a complete model, checks were made on the modeling procedures by replacing the probe and borestem by lunar material containing either: 1) a spherical surface source,, or 2) an infinite line source. Using computer models representing a lunar medium with these two sources, transient temperature distribution within the lunar medium were calculated and compared with available exact solutions for the same physical problem. These studies and comparisons verified that the computer modeling procedures developed for the lunar medium were accurate and could be used

Table 2   Summary of parameters for modeling
lunar surroundings of probe

Quantity	Mode 2	Mode 3
Length of vertical subdivision (cm)	1.091	1.846
Number of vertical subdivisions	13	15
Number of radial subdivisions	5	6
C, ratio of outer-to-inner radius of each annulus	1.367	1.512
Number of finite-difference heat flow equations describing lunar medium	213	291
r_o, inner radius of the borestem	1.256	1.256

in conjunction with the models of the probe to simulate lunar
operation of the conductivity measurements.

As a final step the numerical model of the probe, the borestem and the surrounding medium are combined to make a complete
model of Mode 2 and Mode 3 operation for each experiment location.

Theoretical Predictions for Performance in Lunar Environment

Mode 2 Results

In Fig. 5a the temperature rise at location 2 predicted by
the numerical analysis for four thermal conductivities is shown.
The temperature rise of the probe radiating to a black cylindrical cavity is also shown for comparison.   It is apparent from
these results that the sensitivity of the gradient sensor temperature rise to conductivity becomes very small for values of
k greater than 4×10^{-4} W/cm-K.

The rate of temperature rise as a function of temperature
is shown in Fig. 5b.   About 4 hrs after heater turn-on, the
slope vs time plot on a log-log scale is nearly a straight line

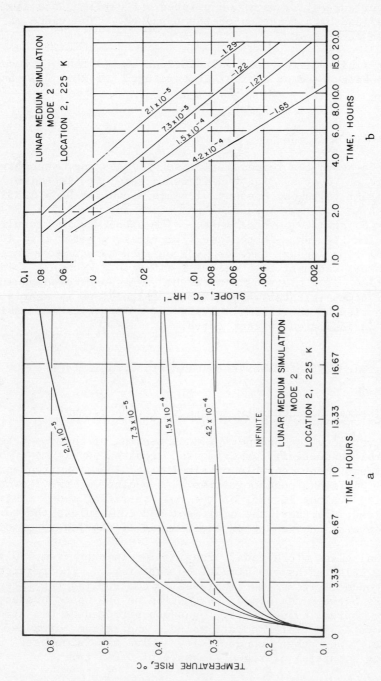

Fig. 5 a) Temperature rise vs time calculated using the finite-difference models for four conductivity values, shown on the curves. The curve labelled "infinite" shows the temperature rise in an isothermal, cylindrical cavity.
b) The slope vs time for four conductivities. The negative members near the bottom of the curves are the values of β (see p. 18).

for the four curves. Thus as we pointed out earlier, the slope
is well defined by a function of the form Ct^β. The value of β
is given on each curve.

Similar finite-difference calculations at all four locations have been made at ambient temperatures of 205, 225 and
245 K.

Mode 3 Results

Figure 6a shows the temperature rise of the ring sensor vs
time predicted by the finite-difference model at location 2.
Five different values of conductivity are shown for $\rho c = 1.339$
W-sec/cm^3-K as well as an additional case for $k = 1.7 \times 10^{-3}$
W/cm-K but ρc at 1.607 W-sec/cm^3-K. The slope vs time results
are shown in Fig. 6b. Note that for times greater than 100
minutes the ring sensor exceeds its designed temperature range
for the lowest conductivity case $k = 1.5 \times 10^{-4}$ W/cm-K. The
slope vs time also plots as a straight line on the log-log
scale for times greater than 240 min. The slope is nearly a
linear function of the thermal parameter. The value of this
parameter is indicated on each curve.

Development of Laboratory Simulation Facilities

Variable Thermal Conductivity Apparatus (K apparatus)

Design verification and performance testing of the lunar heat
flow thermal conductivity experiments required development of
laboratory equipment capable of simulating lunar environments
of several different conductivities. As a minimum, two thermal
conductivity values in each of the two operating mode ranges
were required. Further, the equipment had to simulate the temperature levels and the vacuum environment of the lunar subsoil.

Spherical glass microbeads, ranging in diameter from 590 to
840 μ were selected as the simulation material. The required
range of conductivities was achieved by controlling the inter-
stitial gas between the beads; evacuation of the space produced
the lowest thermal conductivity, gaseous nitrogen at one atmosphere produced an intermediate value, and helium produced a
high value. One other conductivity value was simulated by
maintaining a partial nitrogen pressure of 0.045 torr.

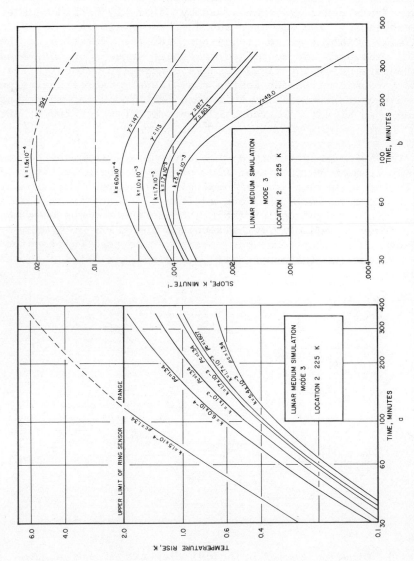

Fig. 6   a) Temperature rise vs time for Mode 3 finite-difference models. Note that one case ρc is increased by 20% to 1.607 W-sec/cm3-K.   b) Slope vs time for Mode 3. Note: Units for gamma given on curves are in cal-1-sec-1/2-cm2-K.

The test apparatus illustrated in Fig. 7 consists of two interconnected chambers to allow testing of a pair of heat flow probes in a vacuum environment.   The probes are inserted into thin-walled epoxy fiber glass sleeves, having a 2.5 cm i.d.. The sleeves are evacuated to less than 5×10^{-5} torr during tests.   The sleeves are surrounded by carefully packed beds of the microbeads.   Bed diameter and depth were chosen to provide essentially "infinite" surroundings for the duration of the tests.   The beds are contained in closed vessels so that bead pressure outside of the sleeve can be controlled without affecting the vacuum environment inside the sleeve.   A refrigeration coil mounted to the wall of the bead tank cools and maintains the bead bed at any desired temperature between 200 K and 250 K.

The thermal conductivity of the evacuated glass beads was shown by tests with the heat flow probes to be a function of depth in the tank and of residual gas pressure.   Tests at Arthur D. Little, Inc. (ADL) and additional tests at the NASA Marshall Space Flight Center (MSFC) showed the evacuated glass beads to have a conductivity between 0.3 and 2.1×10^{-4} W/cm-K depending on the residual gas pressure and the compressive load.   These measurements are discussed in detail in the Appendix.

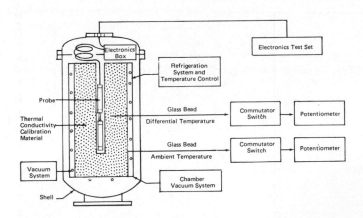

Fig. 7 Diagram showing principal elements of one of the conductivity test chambers (K apparatus). Two such chambers were made so that two probes could be tested simultaneously.

The glass bead beds with the void space filled with N_2 and He at one atmosphere and a temperature of 225 K had conductivities of 1.72×10^{-3} and 5×10^{-3} W/cm-K, respectively.

Tests in an Isothermal Environment

Temperature measurement performance tests of the heat-flow probes required development of another thermal test apparatus, the "gradient apparatus," to provide controlled ambient temperatures and gradients. The test section of this apparatus contains a temperature controlled, thick-walled aluminum tube. The inner wall of this tube is blackened. The tube can be operated with gradients less than 0.1 K over a 50 cm length, simulating an isothermal environment of infinite conductivity. Mode 2 tests were run in this apparatus to give data on the thermal characteristics of the probe at each experiment location. Results of these tests were used to verify the finite-difference model of each probe section.

Test Procedures

The tests are performed with the complete heat-flow subsystem, i.e., two heat flow probes connected to the electronic system. Each of the probes is placed in one of the two K apparatus tanks, or for tests in the gradient apparatus, one probe is placed inside the controlled gradient tube. The bead bed and sleeve are equilibrated at the desired temperature and monitored to assure that the temperature is stable. If tests are being made in evacuated glass beads, the equilibration is accelerated using gaseous nitrogen in the tank to increase conductivity. After equilibration, the tank is evacuated to a level of about 0.001 torr.

The heater at the location being tested is energized in either Mode 2 or Mode 3 and appropriate data recorded continuously during the time the heater is on. For Mode 3 tests data were also taken for a 24 hr period after the heater was turned off. The data were reduced using the same procedures to be used for lunar data.

Mode 2 Tests in the Gradient Apparatus

Table 3 compares the gradient sensor temperature rise at steady state predicted by the finite-difference models of the four locations with corresponding test results from five heat flow probes. The predicted value always falls within the

Table 3   Comparison between measured and predicted results
for Mode 2 performance in isothermal environment

Location	SN-3 Probe 1	SN-4 Probe 1	SN-4 Probe 2	SN-6 Probe 1	SN-6 Probe 2	Finite-difference
1	0.197	0.203	0.212	0.200	0.197	0.200
2	0.206	0.212	0.202	0.221	0.214	0.213
3	0.217	0.215	0.217	0.215	0.217	0.216
4	0.225	0.212	0.218	0.211	0.224	0.223
2 (205 K)	0.264					0.266
2 (245 K)	0.160					0.170

scatter of the test results indicating that the thermal performance of the probe is adequately modeled. The variation between
probes is due to slight differences in constructional detail.
These differences, which are as large as 0.012 K in one case,
must be taken into account when interpreting the data using the
finite-difference models. A comparison between predictions for
the transient temperature response and test results is shown in
Fig. 8 for two experiment locations. The agreement between experiment and theory is good over the total time to reach steady
conditions.

Mode 2 Tests in the K Apparatus

Temperature rise vs time data for Mode 2 tests in the K
apparatus are shown in Fig. 9a, along with data from the gradient apparatus tests. These tests were made on one of the qual-
ification units (HFE SN-3). Note that for tests in the K apparatus at 0.6 μ and 225 K the temperature rise at location 1 is
highest, whereas that at location 4 is lowest. This order is
exactly opposite to results in the gradient apparatus, and indicates that the conductivity of the bead bed increases with
depth sufficiently to overcome locational variation. Secondly,

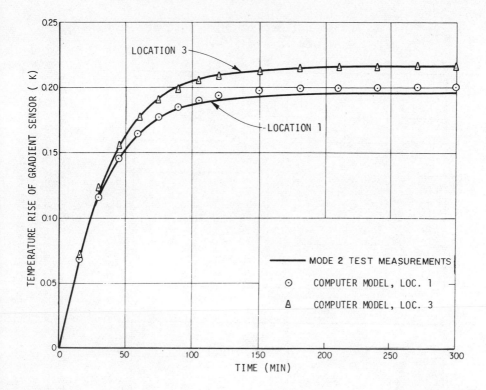

Fig. 8   Transient response of gradient sensor for Mode 2
tests in isothermal cylindrical cavity compared with predicted
results.

the temperature rise during the test at 205 K shows the pronounced dependence on temperature.   Note that the difference
between the 205 K and 225 K temperature rise for k tests remains constant after several hours and the difference is nearly the same as that obtained in the gradient apparatus tests.

    A finite-difference model of the probe in the K apparatus
was made for location 2, Mode 2.   Computations were made for
conductivities of 7.3 x 10^{-5} and 1.5 x 10^{-4} W/cm-K. For these
models ρc was set at 1.256 W-sec/cm^3-K, the best estimate for
the glass microbeads.   These theoretical curves are shown in
Fig. 9 for a 225 K temperature. They indicate an experimental
value for thermal conductivity at location 2 of slightly more
than 1.5 x 10^{-4} W/cm-K.

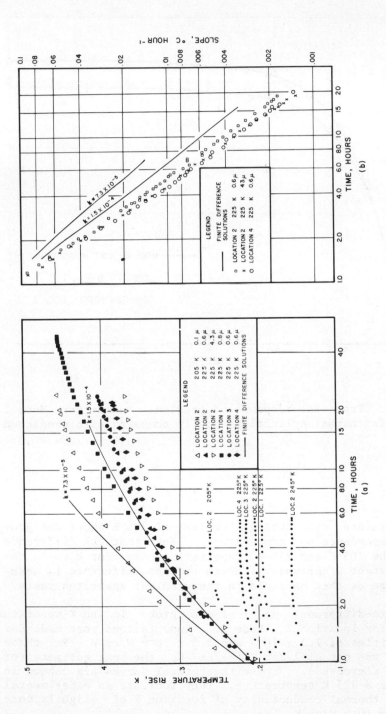

Fig. 9 a) Temperature rise vs time for tests in the K apparatus and gradient apparatus. Finite-difference calculations using a model of the K apparatus with bead bed conductivities of 7.3 x 10⁻⁵ and 1.5 x 10⁻⁴ W/cm-K. Small dots show data from gradient apparatus tests. b) Slopes vs time for some of the tests compared with finite-difference results.

In Fig. 9b we also show the slope $\partial T/\partial t$ vs time for several tests run in the K apparatus and theoretical results.   Note that on this log-log plot the data nearly describe a straight line as predicted by the elementary theory presented earlier.

In Fig. 10 the conductivity deduced from eight different Mode 2 tests at location 2 and one at location 3 are plotted vs N_2 gas pressure in the bead bed.   The points indicate an in- crease in conductivity with gas pressure, which is discussed more fully in the Appendix. Plotted on the same figure are re- sults of conductivity measurements made at MSFC.   There is good agreement in the trend of conductivity with gas pressure be- tween the MSFC measurements and the probe results. The compres- sive load on the beads in the K apparatus at location 2 is about 0.15 kg/cm^2.   The MSFC results lie systematically below the probe results despite the higher compressive load.   In the Appendix we give evidence that the number of interparticle

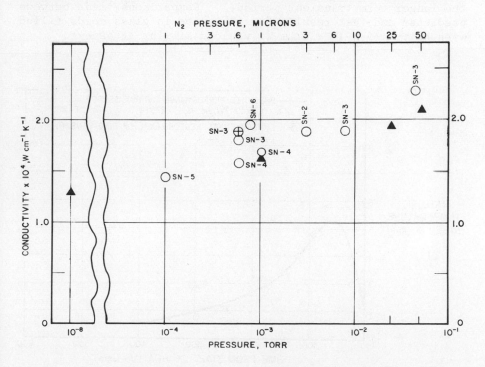

Fig. 10 The deduced conductivity of the glass bead bed from Mode 2 tests plotted vs N_2 pressure. Open circles show loca- tion 2 results and the crossed circle shows location 3 results. Solid triangles show results of independent measurements of con- ductivity of the glass beads at MSFC under a compressive load of 0.35 kg cm^{-2}.

contacts play a large role in determining the absolute value
of the thermal conductivity; thus this difference may be real
and not a result of systematic measurement error.   However,
the tests to date do not giv. us a clear basis for decision.
In situ, line source measurements in the K apparatus bead beds
are planned to resolve the difference.

Comparison of Mode 3 Tests with Predictions

    Computer predictions for the rate of ring-sensor tempera-
ture rise were compared with results of tests performed in the
K apparatus. Figure 11 shows predicted and measured values
for rate of sensor temperature rise vs time when the bead bed
surrounding the probe was filled with one atmosphere of gaseous
nitrogen (k = 1.72 x 10^{-3} W/cm-K).   Close agreement is shown
between computer-predicted and test results especially during
the longer term transient period.   Comparisons made between
predicted and test results for the probe in glass beads filled
with helium (5 x 10^{-3} W/cm-K) yielded similar agreement.

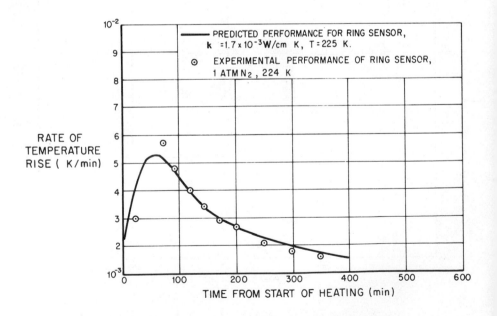

    Fig. 11   Comparison of predicted and test results for
Mode 3 experiment in K apparatus.

Procedure for Making Lunar Conductivity Experiments
and Analysis of the Data

The in situ measurements will begin approximately two and
a half weeks after the probes are implanted in the lunar soil.
This will allow about 400 hrs for the probes to equilibrate
with the lunar subsurface. During this period continuous data
from all gradient bridges and thermocouples will be recorded.
Daily measurements of the ring bridge temperature will also be
made.

An initial estimate of the soil conductivity can be made
based on the equilibration time of the probes and the magnitude
of the thermal waves detected by the thermocouples. It is possible that the probes will not have fully equilibrated at the
end of the 400 hr period. However, the change in temperature
should be linear enough to allow accurate extrapolation over a
40 hr period. Before a measurement is started, we will examine
100 hrs of temperature data to establish the temperature trends.

The Mode 2 measurements will be performed first. Each heater
will be turned on for at least 36 hrs and the eight experiments will be run in a sequence that causes the least thermal
disturbance to the subsequent measurement. Thereafter, a
similar sequence will be followed for the Mode 3 measurements,
with each measurement requiring approximately 20 hrs. The heater will be energized for approximately 8 hrs; after the heater
is turned off, the decay of temperature of the gradient sensor
beneath the heater and adjacent sensors will be monitored for
about 12 hrs. All of the experiments will have been completed
about 1000 hrs after the heat flow experiment was emplaced.

We will depend principally on the finite-difference models
to interpret the data from the experiments. A model has been
made for each location and we have used them to establish the
relation of temperature rise and slope to thermal conductivity
at various times after heater turn-on for both modes.

Figure 12 shows a graphical summary of such relations for
location 2 at an ambient temperature of 225 K. On the left
is an example of temperature rise vs conductivity and slope vs
conductivity for Mode 2, 16 hrs after heater turn-on. Note
that on the log-log scale temperature rise vs conductivity very
nearly plots as a straight line. Results at other ambient temperatures form lines parallel to the 225 K line. The relation
between slope and conductivity assumes a value of ρc equal to
1.34 W-sec/cm^3-K.

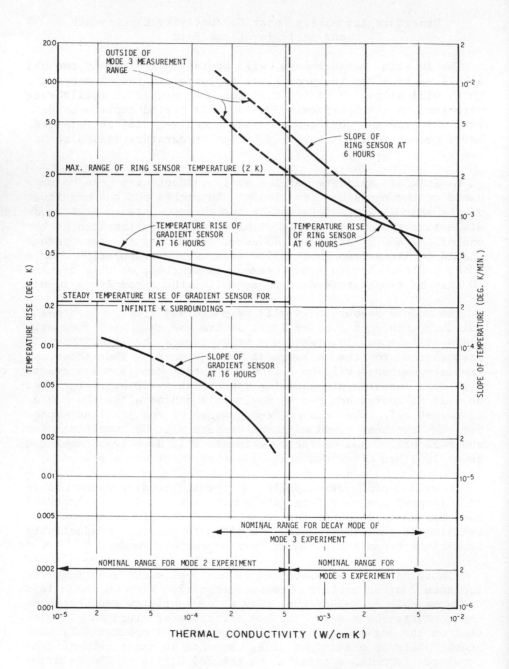

Fig. 12 Summary diagram showing temperature rise (left-hand ordinate) and rate of temperature rise (right-hand ordinate) vs thermal conductivity. $\rho c = 1.339$ W-sec cm^{-3} K^{-1}.

On the right or high end of the conductivity range similar relations for Mode 3 are shown.   Note that for conductivities less than 5 x 10^{-4} W/cm-K the ring sensor exceeds its designed range and this sets the lower limit for Mode 3 measurements. The decay mode refers to interpretation of data as the sensors are reequilibrating with the lunar subsurface after the heater in Mode 3 has been turned off. This mode provides an important check on both experiments in the range 1.5 x 10^{-4} to 5 x 10^{-3} W/cm-K.

The finite-difference models have been used to produce tables giving the relation for all locations at several designated times after heater turn-on. These tables will be used to make preliminary estimates of the conductivity as the data are received. One of the primary requirements for such rapid anal- ysis is to aid in running the measurements in the most effi- cient and accurate manner.

The finite-difference models will be used to refine the pre- liminary estimates by varying the simulated lunar medium con- ductivity until the observed response is reproduced.   In some cases this may involve assuming the lunar medium is heteroge- neous. The multiplicity of measurement locations should provide a clear indication of local heterogeneities.

Discussion of Errors in the Analysis of Experiments

Since we will primarily use the finite-difference models to analyze the data from the lunar conductivity measurements, questions arise involving their accuracy and possible errors associated with their use. Also of importance in interpreting conductivity measurements is the relationship between an error in probe sensor response and the deduced value of conductivity.

The available data indicate that the finite-difference models adequately describe the thermal performance of the probes. Also uncertainty analyses performed to date indicate that use of finite-difference models probably will not be a limiting factor in the interpretation of lunar conductivity data. However, the influence of uncertainties in the mathemathical models of the probe are a function of the conductivity of surrounding materi- al; and therefore, a refined error analysis is planned after actual results are obtained from the moon.

The fact that small changes in predicted temperature rise result when large values of parametric uncertainties are included in the mathematical models is also favorable. In Table 4 we show the result of a study of how uncertainties in the assignment of certain thermal parameters affect the temperature rise of a Mode 2 measurement in an infinite-conductance surrounding. Some of the specified uncertainties are quite large and have a small influence on the temperature rise. The most important parameter of those listed is the emittance of the probe body. The two-sigma variation in temperature rise considering all the uncertainties shown in Table 4 is less than 3%.

Table 4   Influence of uncertainties on predicted temperature rises in the isothermal cylindrical cavity

Parameter	Change, %	Change in ΔT, %
1) Thermal conductance or probe body	Increase 10	−0.7
2) Lead wire conductance	Decrease 21	+0.7
3) View area from heater to space	Increase 5	−1.1
4) Emittance of sensor can	Decrease 43	+0.2
5) Emittance of probe body	Decrease 3	+2.5
RMS:		2.9

The relation between errors in temperature rise and errors in deduced conductivity is a function of conductivity of the surrounding medium and the mode of operation. For the Mode 2 experiments, on the average, a 1% error in temperature rise results in a 4% error in conductivity. Thus to meet our preset error limit of 20%, the highest permissible error in temperature rise prediction is 5%. Using the slope data to analyze Mode 2 data offers some advantages since the slope is more sensitive than temperature rise to changes in conductivity (see Eqs. 1 and 2). In addition the slope is less sensitive to uncertainties in modeling of probe details. However these advantages are offset by problems in estimating ρc and in the intrinsic errors in measuring the slope accurately.

For the Mode 3 measurements, a 1% error in slope results in a 1% error in conductivity, a more favorable relationship between error in probe response and deduced conductivity than that which occurs in the Mode 2 measurements. An illustration of the low sensitivity of the Mode 3 measurement to small uncertainties in probe thermal parameters is that the predicted slopes at four experiment locations varied by less than 3%, when the lunar medium conductivity was 1.7×10^{-3} W/cm-k. Also, a 20% uncertainty in the thermal conductance of the borestem caused a 2% error in slope prediction. The errors in reducing data in the decay mode are about the same as those for Mode 3.

APPENDIX

Thermal Conductivity of Evacuated Glass Beads

Evacuated Glass Beads

Measurements of glass microbeads used in the K apparatus under evacuated conditions were made in the guarded cold-plate apparatus at Arthur D. Little, Inc.. Several measurements were made in temperature range of 200 to 250 K at three compressive loads up to 0.34 kg/cm^2. The best value determined from these experiments at 225 K is 7.3×10^{-5} W/cm-K. Watson[8] measured the conductivity of microbeads in the same size range, .590 to .840 mm. At 225 K, the mean value determined by Watson was about 3×10^{-5} W/cm-K; approximately one-half the above value.

The guarded cold-plate measurements indicate no detectable changes of conductivity with compressive load. However, the conductivity tests made with the heat-flow probes in the K apparatus when interpreted using the finite-difference models showed that; 1) the thermal conductivity of beads in the tank is greater than either that measured in the guarded cold-plate apparatus or by Watson, and 2) there is a vertical gradient in conductivity in the bead bed. The existence of a conductivity gradient implies a relation between compressive load and heat conduction through the beads. Consequently, further conductivity measurements were made at the NASA Marshall Space Flight Center.

The measurements at Marshall were made using a line heat-source technique and the analysis technique used the time derivative instead of the more customarily used temperature vs time curves, Merrill[5]. This method of analysis requires only a short observation time, 0.5 to 1.0 minutes, and also provides useful results even when the ambient temperature of the bead bed is increasing uniformly with time. Measurements were made on beads with no load at very high vacuum, 10^{-8} torr, and with compressive loads of 0.198 and 0.35 kg/cm^2. The results of these measurements are shown in Fig. 13. Numerous measurements were made over the range 190 to 300 K. The curves drawn through the data is a best fit to the function $k(T) = B + AT^3$ after Watson[8] and the scatter in terms of the standard deviation about the curves, which is quite large, is given in the figure. The best value for conductivity before any load is applied at 225 K is 3.0 x 10^{-5} W/cm-K; very near to the value obtained by Watson.

As compressive loads were added to the bead bed and the tests repeated, the conductivity increased by nearly a factor of five (see Fig. 13). Also interesting is the fact that when the load is removed completely without disturbing the beads (open circles in Fig. 13) the measured conductivity at 300 K is well above those measured before the load was applied. These observations imply that the conductance across contacts between grains is important. It seems that loading the sample results in a greater number of interparticle contacts than when the beads are loosely loaded and that these contacts are partially preserved even after the load is removed.

Why the tests under load in the guarded cold-plate apparatus failed to show a dependence on compressive load remains unresolved. We suspect that bridging within the sample may have prevented the stress from being transmitted throughout the sample.

When the K apparatus bead beds were loaded, great care was taken to compact the bed uniformly to a density of 1.59 g/cm^3 (35% voids) by repeated short drops of the container. If the beads in the tank will support no shearing stress, the weight of the beads above experiment #1 would correspond to a load of

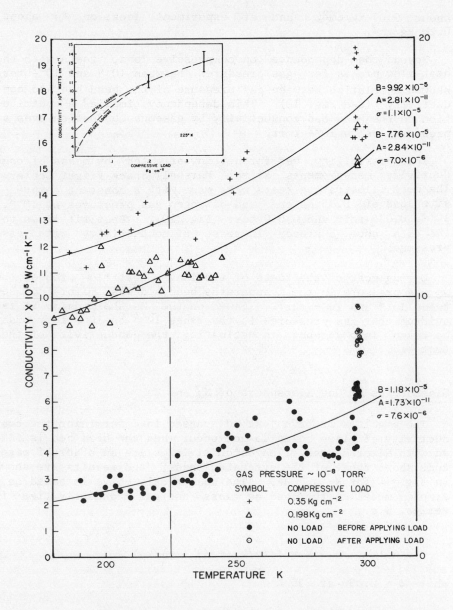

Fig. 13 Results of conductivity measurements on glass beads
at MSFC for various compressive loads.   The values of A and B
correspond to coefficients in the equation $k(T) = B + AT^3$.   The
inset in the upper left shows conductivity vs load at 225 K
with two possible trends suggested.   The glass beads were loaded in the K apparatus as received, therefore the beads were not
specially cleaned for the tests at MSFC.

about 0.07 kg/cm^2,   and at experiment location #4 about
0.28 kg/cm^2.

Beyond the dependence on compressive load, tests with the
heat-flow probes for gas pressures between 10^{-4} and 10^{-2} torr
showed a relation between gas pressure of the bead bed and conductivity (see Fig. 10).   This dependency implies a contribution to the bead bed conductivity by gaseous conduction even at
pressures below 10^{-2} torr.

This possibility was further investigated by means of conductivity measurements at the Marshall Space Flight Center.
The most illustrative tests were made with a constant compressive load of 0.35 kg/cm^2 and nitrogen gas pressures of 10^{-8},
10^{-3}, 0.025 torr and 0.050 torr (Fig. 10).   Best fit lines to
the data show a steady increase of conductivity with gas
pressure.

During most of the tests of the lunar heat flow probes, the
N$_2$ gas pressure in the K apparatus bead bed was maintained between 10^{-4} and 10^{-2} torr.   Based on the Marshall data, it is
evident that gas pressures in the range 10^{-4} to 10^{-2} torr must
be taken into account in estimating the conductivity of the
test bed.

Glass Beads at One Atmosphere of N$_2$ and He

The problems we have just discussed in determining the conductivity of glass beads do not occur when the bead bed is filled with N$_2$ or He gas. The conductivities are an order of magnitude above those of the evacuated beads, the results are shown
in Fig. 14 for N$_2$ and He.   The analytical curve is based on a
simple model of the bed as glass and gas conductivities in
series, i.e.,

$$k = 1/(\phi/k_g + (1 - \phi)/k_s)$$

where ϕ = 0.096 at 230 K

k_g = conductivity of the gas

k_s = conductivity of the glass.

The best values at 225 K are 1.72 x 10^{-3} and 5 x 10^{-3} W/cm-K
for N$_2$ and He filled beads respectively.

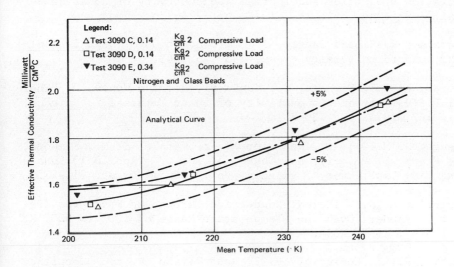

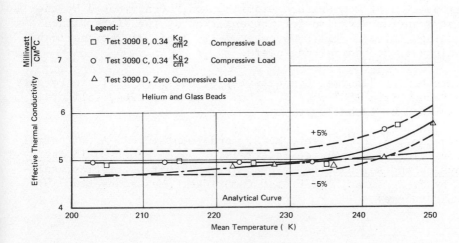

Fig. 14 Top: Results of conductivity measurements in
glass beads filled with 1 atm of nitrogen are plotted vs temperature. Bottom: Results of measurements with 1 atm of
helium.

References

[1]Birkebak, R. C., Cremers, C. J., and Dawson, J. P., "Thermal Radiation Properties and Thermal Conductivity of Lunar Material," Science, Vol. 167, 1970, pp. 724-726.

[2]Carslaw, H. S. and Jaeger, J. C., Conduction of Heat in Solids, Oxford University Press, 1959, p. 263.

[3]Horai, Ki-iti, Simmons, G., Kanamori, H., and Wones, D.,"Thermal Diffusivity and Conductivity of Lunar Material," Science, Vol. 167, 1970, pp. 730-731.

[4]Langseth, M. G., Jr., Wechsler, A. E., Drake, E. M., Simmons, G., Clark, S. P., Jr., and Chute, J., Jr., "Apollo 13 Lunar Heat Flow Experiment," Science, Vol. 168, 1970, pp. 211-217.

[5]Merrill, R. B., "Thermal Conduction Through an Evacuated Idealized Powder Over the Temperature Range of 100° to 500° K," TN D-5063, 1969, NASA.

[6]Robie, R. A., Hemingway, B. S., and Wilson, W. H., "Specific Heats of Lunar Surface Materials from 90 to 350°K," Science, Vol. 167, 1970, pp. 749-750.

[7]Strong, P. F. and Emslie, A. G., "The Method of Zones for the Calculation of Temperature Distribution," Contributed by the Heat Transfer Division for presentation at the Winter Annual Meeting, Chicago, Ill., November 7-11, 1965, of The American Society of Mechanical Engineers.

[8]Watson, K., "Thermal Conductivity of Selected Silicate Powders in Vacuum from 150°-350° K," Ph. D. thesis, 1964, California Institute of Technology.

[9]Wechsler, A. E., Glaser, P. E., and Fountain, J. A., "Thermal Properties of Granulated Materials," AIAA Progress in Astrophysics and Aeronautics: Lunar Thermal Characteristics, edited by J. W. Lucas, MIT Press, Cambridge, Massachusetts, 1971 (this volume).

THE APOLLO 15 LUNAR HEAT FLOW MEASUREMENT

Marcus G. Langseth Jr.*

Lamont-Doherty Geological Observatory of Columbia University,
Palisades, N. Y.

Sydney P. Clark Jr.+

Yale University, New Haven, Conn.

and

John Chute Jr.‡ and Stephen Keihm§

Lamont-Doherty Geological Observatory of Columbia University,
Palisades, N. Y.

The lunar heat flow experiment (HFE) described in Chap.
2c was successfully emplaced on Apollo 15.[1] The relatively
high density of the regolith at Hadley Rille prevented full
penetration of either of the two fiberglass borestems to the
planned 3-m depth. Both borestems were driven about 1.6 m
into the subsurface; however, one of the probes, designated
#2, could not pass to the bottom of the borestem because of
an obstruction. The bottom of this probe is only 1 m below
the surface. Figure 1 shows the geometry of the heat flow
probes at the Hadley Rille site.

Data received during the first $1\frac{1}{2}$ months after the
probes were emplaced have been analyzed. During this period
the initially warmer probe and borestem cooled slowly toward
the initial undisturbed temperature distribution in the rego-

Editorial Note: This brief paper on the latest results
obtained from the Apollo 15 flight to the moon, during which
an apparatus of the type described in Chap. 2c was emplaced,
was prepared for this volume just as it was going to press.
*Senior Research Associate.
+Professor of Geophysics, Dept. of Geology.
‡Research Associate.
§Research Assistant.

lith. The mean temperature at 100 cm below the surface is
252.3°K at probe #1 and 250.7°K at probe #2. These values
are nearly 30°K higher than the average surface temperature
over a lunation cycle, which is about 217°K. Most of this increase in mean temperature is the result of the high dependence of heat transfer on ambient temperature in the upper
several cm of the lunar soil.[2] The temperature data from
probe #1, which was emplaced at 18 hr, 47 min GMT, on July
31, 1971, for the first 300 hr of operation are shown in Fig.
2.

Eight temperature sensors below 70 cm (6 from probe #1
and 2 from probe #2) were unaffected by the large variations
of surface temperature during the first 30 days after emplacement. The cooling histories of these thermometers were
extrapolated to their equilibrium values using the theory of
cooling of a cylinder in an infinite medium. After applying
appropriate corrections for the shunting effect of the probe
and corestem, these extrapolated temperatures yield an accurate determination of the temperature field that existed in
the subsurface before the probes were inserted. This analysis indicated that the undisturbed vertical temperature gradient at probe #1 below 70 cm is 1.75 $\pm$ 0.09°K/m.

Each of the heat flow probes has heaters surrounding
four of the thermometers. These heaters can be turned on by
command from Earth to conduct in situ experiments (described more fully in Chap. 2c) to determine the thermal
properties of the soil surrounding the borestem. Six such
experiments were conducted in August and September. The
analysis of these experiments indicated that the conductivity
of the regolith is not uniform with depth. The values obtained from three experiments (1 with probe #2 and 2 with
probe #1) are given in Table 1.

Table 1 Lunar material thermal conductivity values

Temperature sensors		
No.	Depth, cm	Conductivity,[a] W/cm-°K
TG 22A	49	1.4×10^{-4}
TG 12A	91	1.7×10^{-4}
TG 12B	138	2.5×10^{-4}

[a]Estimated error of these measurements is $\pm$ 10%.

Temperature rise data and theoretical curves fitted to the
data are shown in Fig. 3. The theoretical curves are computed, using the finite-difference models described in Chap.
2c.

An independent estimate of thermal conductivity can be made from the cool-down history of the thermometers. Such estimates are dependent on knowledge of the initial temperature and consequently are not very accurate ($\pm 50\%$). Analysis of the first 100 hr of data indicate that the conductivity increases with depth at a rate similar to that shown by in situ measurements utilizing heaters. The value at 91 cm is 1.5 x 10^{-4} ± 50%, and the value at 138 cm is 2.7 x 10^{-4} ± 50%. These values are in good agreement with the in situ measurements utilizing heaters. (Also see Chap. 3b for comparison with measurements on lunar material brought to Earth by previous Apollo missions.)

The conductivity data, combined with the gradient determination, indicate that the average heat flow through the regolith is 3.3 x 10^{-6} W/cm² (0.79 x 10^{-6} cal/cm² sec) at the probe #1 site. For comparison, this value is about ½ the average heat flow from Earth.

The shallow emplacement of probe #2 prevents an accurate estimate of the mean gradient with only 1000 hr of data. When several lunations of data become available, the heat flow at this site can be calculated more accurately.

Because of the proximity of the heat flow measurement to the surface, it is susceptible to a number of disturbances. The principal ones are caused by topography and refraction of heat flow associated with sloping interfaces between materials of differing thermal conductivity. The effects of the most important topographic features, Hadley Rille and the Apennine Front, have been estimated by modeling them with very simple geometries. These preliminary estimates indicate that positive corrections on the order of 10 to 20% should be applied to the observed heat flow value.

Heat flow measurements are planned for Apollo 16 and 17. Realizing that these future measurements may produce major changes in our conclusions, we consider briefly the implications of the observed heat flow, taking it at face value.

For a planetary body the size of the moon, very little of the present heat flow results from initial temperatures even if they were at the solidus. Most of the present heat flow must result from the decay of long-lived radioisotopes. Isotopic abundances appropriate for ordinary chondrites would result in present rates of heat loss of 1.0 to 2.0 x 10^{-6} W/cm².[3] A moonwide heat flow of 3.3 x 10^{-6} W/cm² would require much higher rates of heat production at present than a "chondritic" bulk composition for the moon can provide.

References

[1]Langseth, M. G., Jr., Clark, S. P., Jr., Chute, J. L., Jr., Keihm, S. J., and Wechsler, A. E., "Lunar Heat Flow Experiment," APOLLO 15 Preliminary Science Report, NASA, Washington, D. C. (in press).

[2]Linsky, J. L., "Models of the Lunar Surface Including Temperature Dependent Thermal Properties," Icarus, Vol. 5, 1966, pp. 606-634.

[3]Fricker, P. E., Reynolds, R. T., and Summers, A. L., "On the Thermal History of the Moon," Journal of Geophysical Research, Vol. 72, 1967, pp. 2649-2661.

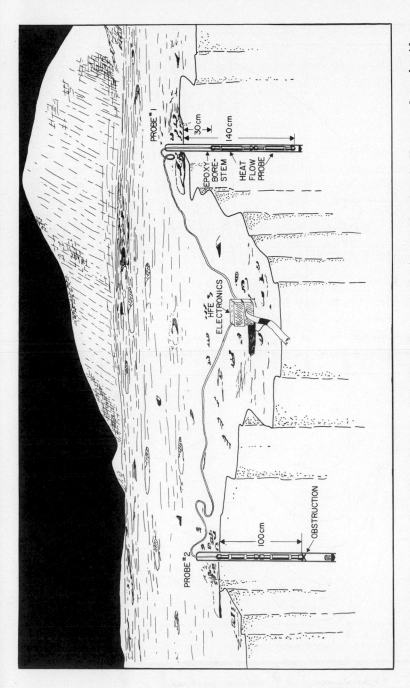

Fig. 1 A cutaway drawing, looking north, showing the positions of the heat flow probes in the lunar soil at Hadley Rille Base. The probes are emplaced about 8.5 m apart. The electronics box, which houses the detecting, programming, and signal conditioning circuits for the experiment, sits on the surface midway between the two probes.

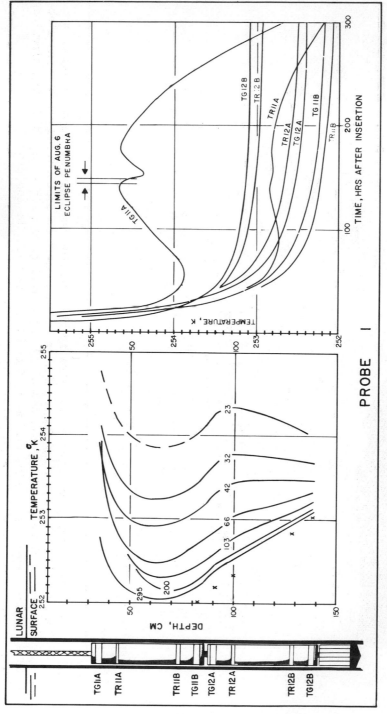

Fig. 2   Temperature data from probe #1 during the first 300 hr of operation. The position of the eight thermometers below the lunar surface is shown on the left. The graph on the left shows the temperature profiles along the probe for various times (given in hours on each curve) after emplacement. The graph on the right shows the same data plotted vs time.

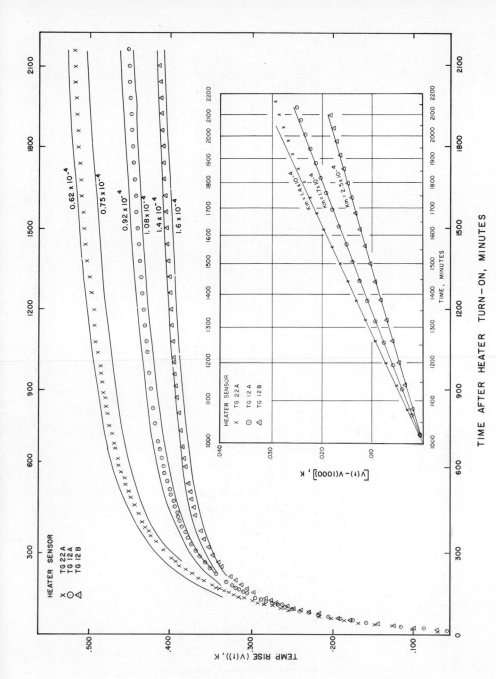

TIME AFTER HEATER TURN-ON, MINUTES

Caption for figure #3 appears on following page.

Figure #3 appears on previous page.

Fig. 3 Temperature rise data for three conductivity experiments. The main plot shows the data on a linear scale. The inset shows temperature rise minus the temperature rise at 1000 min plotted on an expanded logarithmic time scale. The solid lines in the inset are theoretical curves derived from finite difference models best fitted to the data. The conductivity in $W/cm-^{\circ}K$ of the surrounding lunar medium (Km) for each model is shown on the curve. The solid curves on the linear plot are theoretical curves using the same Km's. The thermal contact conductance $W/cm^2-^{\circ}K$, between the borestem and the lunar medium, has been adjusted for best fit. Curves corresponding to bracketing models are given, and values of contact conductance are shown. However, a more exact value of contact conductance can be obtained by interpolation.

SECTION 3. THERMAL CHARACTERISTICS OF LUNAR-TYPE MATERIALS

THERMAL PROPERTIES OF GRANULATED MATERIALS

Alfred E. Wechsler* and Peter E. Glaser[/]
Arthur D. Little, Inc., Cambridge, Mass.

and

James A. Fountain[#]
NASA George C. Marshall Space Flight Center, Huntsville, Ala.

I. Introduction

Prior to the Surveyor and Apollo lunar landings, lunar surface materials had been characterized by earth-based observation techniques supplemented by laboratory experiments on possible constituents of the lunar surface. The observations were based on passive measurements--photometry, polarization, infrared, and radio wave emission--and active measurements--microwave and radar (see Chapters 1.a, 1.b and 1.c, respectively). Laboratory studies were generally confined to determining the optical, thermal, and dielectric properties of porous and powdered rocks and soils at high vacuum conditions designed to simulate the lunar surface environment.

Consideration of the thermal balance of the lunar surface led early investigators to conclude that the variation in surface temperature (or, more correctly, the radiation flux emitted by the surface) during lunations and eclipses depended primarily on the thermal parameter $(k\rho c_p)^{-1/2}$ or, alternatively, the thermal inertia $(k\rho c_p)^{+1/2}$ where k is the thermal conductivity (g-cal/sec cm K), ρ is the bulk or apparent density (g/cm^3), and c_p is the specific heat (g-cal/g K). Wesselink,[1] Jaeger and Harper,[2] and Jaeger[3] compared measured lunar surface temperature variations (infrared measurements) with analytical and numerical solutions of the energy balance at the lunar surface. From these comparisons, they obtained thermal parameter values of 600 to 1500 (cm^2 K sec$^{1/2}$/g-cal) for the lunar surface layer. On the basis of thermal property data in the literature, it was then concluded that the lunar surface layer was most likely a particulate material of

---

*Group Leader, Engineering Sciences Section.
[/]Head, Engineering Sciences Section.
[#]Member, Space Sciences Laboratory.

relatively low density or a low-density porous or vesicular material. As the number and accuracy of infrared observations of lunar temperatures increased, other models of the heat balance were developed by Linsky[4] and Halajian et al.[5] Their modifications took into account the variation of thermal properties with depth and temperature and the possible layering of lunar surface materials. Thus they refined the understanding of the observational data and permitted greater insight into the characteristics of the lunar surface layer.

Examination of the thermal properties of silicates and other minerals shows that the specific heat does not vary more than about a factor of 2 or 3 with either composition or temperature in the range expected on the lunar surface. Similarly, the density of most silicates varies only from about 0.5 g/cm^3 for a loosely packed powder to about 3.0 g/cm^3 for a solid crystalline material. The thermal conductivity of particulate materials, however, can vary by several orders of magnitude depending on gas pressure, particle size and configuration, temperature, etc.

About 1960, a more concentrated study of thermophysical properties of postulated lunar materials was initiated. This study was prompted by the large variation in thermal conductivity of porous and particulate materials, the scarcity of data in the literature on thermal conductivity of powdered and porous rocks under conditions of temperature and gas pressure which adequately simulate the lunar environment, and the corresponding uncertainty about the possible nature of lunar materials, as interpreted from the thermal parameter. Other objectives of this work were to determine if lunar surface materials could be used successfully for thermal protection of man, propellants, or vehicles; to establish the feasibility of economically important processes, such as water extraction on the lunar surface; and to provide complementary information on contact resistance, adhesion, cohesion, and mechanical properties of possible lunar surface materials. It was only after this information had been obtained that the more extreme hypotheses regarding the nature of the lunar surface were discarded and the view accepted that the lunar surface could support the loads of man and his spacecraft.

In this chapter, we review the thermophysical properties of materials believed to simulate the lunar surface layer; our review emphasizes thermal conductivity data and the effects of material and environmental variables on thermal conductivity. Moreover, because of their obvious significance, our discussion is restricted to granular minerals or silicates.

II. Heat-Transfer Mechanisms in Particulate Materials

There are three basic mechanisms of heat transfer in particulate materials: conduction by the gas contained in the void spaces between the particles, conduction within the solid particles and across the interparticle contacts, and thermal radiation within the particles, across the void spaces between particle surfaces, and between void spaces themselves. A common approach to evaluating and interpreting measurements of thermal conductivity of particulates is to assign a contribution of thermal conductivity to each mechanism; these contributions can be summed to give the total or effective conductivity as shown below:

$$k = k_{eff} = k_g + k_s + k_r \qquad (1)$$

A. Gas Conduction

In most particulate materials at room temperature and atmospheric pressure, conduction by the gas filling the void spaces between the particles is the largest contribution to heat transfer. The effects of gas pressure on gas conduction in a particulate material are well known.[6,7] Over a wide range of gas pressures, the thermal conductivity of the gas, and hence the thermal conductivity of the particulate, is independent of gas pressure because the mean free path varies inversely with the pressure and the gas density varies directly with the pressure; the product of mean free path and packing density determines the gas conductivity.

When the gas pressure is reduced to the point where the mean free path within the gas is greater than the interparticle spacing, further reduction in gas pressure does not change the effective free path of the gas. Then the thermal conductivity of the gas in the powder is directly proportional to gas pressure. The pressure at which thermal conductivity becomes proportional to the gas pressure (sometimes called the breakaway pressure) increases as the particle size decreases and as temperature increases. Equations giving the relationship between gas pressure, particle size, and thermal conductivity are available in the literature.[8,9] For powders with particle size from 1 to 200 μ, gas conduction within the powder becomes a negligible heat-transfer mechanism at gas pressures below 10^{-3} to 10^{-2} torr.

Even under conditions where bulk gas conduction is eliminated, residual gas conduction can occur because of "slip-flow heat transfer" or two-dimensional heat flow by molecules adsorbed on

particle surfaces near contact areas.[10] This mechanism could
have been important under lunar pressure conditions if residual
adsorbed layers were not removed by long exposure to high
vacuum or incident solar radiation. Studies of returned lunar
samples indicate that adsorbed gases are not present, and
therefore the residual gas conduction is negligible.

B.   Solid Conduction

In an evacuated particulate material, solid conduction becomes an important heat-transfer mechanism. Heat is transferred within the individual particles by phonon conduction, by
electron conduction, or by thermal radiation, depending upon
the powder type. These processes are the same as in the bulk
solid and can be explained and examined by conventional heat-
transfer theory and measurements.[11] Although studies of contact conduction across macroscopic contacts have been very
successful, it has been difficult to apply these results to a
complex powder system. Information from the theory of electrical contact resistance and thermal contact of metallic joints
has been used to determine the thermal resistance of particles
in contact.[12] Heat transfer at the particle interface across
contact points is difficult to explain and to examine analytically or experimentally because the particle size, shape, composition, density or degree of compaction, and mechanical loading affect the number and the nature of the contact points.
Thus, empirical equations describing contact conduction in the
literature[13,14] may be valid only for certain selected materials. Furthermore, on the moon, the effects of interparticle
forces--electrostatic, Van de Waals, or chemical adhesion--on
contact area may exceed those due to particle shape or mechanical loading. For example, the effect of the reduced gravity on
thermal conductivity was considered by Fremlin[15] in his theory
suggesting the volcanic origin of lunar features.

The unknown effects of the lunar environment made it impossible to predict adequately the heat transfer by solid conduction
in powders, however, and experiments were necessary to clarify
the effects of environmental variables. Nevertheless, several
general conclusions concerning solid conduction heat transfer
can be made: 1) the solid conduction heat transfer in powders
composed of inherently high-conductivity materials (e.g., crystals) will be only slightly higher than in powders composed of
low-conductivity materials (e.g., glasses). This fact was confirmed by experiment many years ago.[6] 2) The greater the
packing density or mechanical loading of the powder, the
greater the solid conduction contribution to heat transfer because of the increased number and area of contacts. 3) Under
high vacuum conditions, where clean surfaces exist, vacuum ad-

hesion could increase solid conduction in nonmetallic powders. The results of experimental measurements which support these conclusions are given later.

C. Thermal Radiation

Thermal radiation is the second important heat-transfer mechanism in evacuated powders. If the materials comprising the powder are opaque to thermal radiation in the wavelength region of interest (usually 4 to 50 μ for powders at temperatures between 100° and 400°K), radiation transport within the individual grains will be small. However, radiation between particle surfaces will still be important.

Empirical and theoretical relations have been derived to predict radiative transport in powders.[16-18] The higher the bulk density of the powder and the smaller the void spaces, the smaller the radiation heat transfer between particles. Radiation heat transfer in an evacuated powder increases with increase in temperature, generally with the third power of the absolute temperature. The magnitude of the radiation contribution can be predicted if the particle size, composition, and optical properties are known. However, the effective optical constants of complex powders are not known, and experimental data are required to evaluate the radiative conductivity.

D. Interaction between Conduction and Radiation

Several investigators[8,9,14,17,18] have examined the interaction between gas and solid conduction and between solid conduction and radiation in powders. In a simple representation of particles in a gas, the effective conductivity of the mixture is a function of the porosity and the ratio of k_s'/k_g (the conductivity of the pure solid, k_s', divided by that of the gas, k_g). As this ratio becomes very large, i.e., $k_g \to 0$, as for high vacuum conditions, unreasonable values of the effective thermal conductivity are often obtained. More complex models give somewhat better results; however, general confirmation with experimental data has been lacking. Fortunately, in examining possible lunar materials in a simulated lunar vacuum environment, the contribution of gas conduction is expected to be negligible. The interaction between radiation and solid conduction is equally complex; although radiation transmission and conduction have been measured, the most useful approach has been to neglect the interaction in reasonably opaque powders and consider the radiation and solid conduction as independent and additive.

Thus the effective thermal conductivity for particulate materials given by Eq. (1) reduces to the following expression for simulated lunar conditions:

$$k_{eff} = k_s + k_r \tag{2}$$

The solid conduction contribution is often a function of temperature, because the strength of the material, and therefore the contact area, can vary with temperature, and because the conductance of the solid itself varies with temperature. The first effect is probably smaller than the second; therefore, the solid conduction term in Eq. (2) can be represented as[16]

$$k_s = B \text{ (invariant with temperature)} \tag{3}$$

or $$k_s = B + CT \text{ (for glasses and silicates whose conductivity increases with temperature)} \tag{4}$$

or $$k_s = B + (D/T) \text{ (for crystalline materials whose conductivity decreases with temperature)} \tag{5}$$

The radiation contribution can be approximated as

$$k_r = AT^3 \tag{6}$$

Thus, the effective thermal conductivity for particulate materials can be expressed as

$$k_{eff} = B + AT^3 \tag{7a}$$

or $$k_{eff} = B + CT + AT^3 \tag{7b}$$

or $$k_{eff} = B + (D/T) + AT^3 \tag{7c}$$

Equation (7a) adequately represents the thermal conductivity of most evacuated particulate materials and has been used by most investigators.

III. Thermal Conductivity of Simulated Lunar Materials

A. Materials Studied

Table 1 summarizes the types of materials, range of environmental variables, and measurement methods used by several

Table 1 Summary of materials used in property measurements

Material studied	Particle size, μ	Temperature, °K	Range of variables Gas pressure, torr	Density, g/cm³	Method[a]	Reference
Basalt	37-62	135-370	10^{-3}-10^{-8}	0.79-1.50	DLHS	26
Basalt	44-104	220-330	10^{-5}-760	1.27	Probe	25
Basalt	10-74	170-360	10^{-5}-10^{-8}	1.36-1.43	LHS	16
Basalt	105-150	280-340	10^{-10}-760	...	LHS	20
Olivine basalt	10-400	170-370	10^{-6}-760	1.14-1.96	Diff	24
Olivine	<74	180-300	10^{-5}-10^{-6}	1.37	Radiative	22
Olivine	<25	260-300	10^{-5}-10^{-10}	...	LHS	20
Pumice	1-150	230-350	10^{-5}-760	0.80-0.87	Probe	25
Pumice	2 cm & fines	300	10^{-2}-760	1.12	Probe	19
Pumice	10-74	160-380	10^{-6}-10^{-8}	0.82-0.84	LHS	16
Hornblende	<74	150-350	10^{-5}-10^{-6}	1.10-1.50	Radiative	22
Tektite	<20	280	10^{-9}	...	LHS	20
Chondrite	<20	225-435	10^{-5}-10^{-10}	...	LHS	20
Expanded perlite	300-350	77-370	10^{-5}-760	0.08-0.092	Cold plate	7
Expanded perlite	300-350	300	10^{-5}-760	0.014	Probe	25
Quartz	44-74	150-350	10^{-5}-10^{-6}	1.10-1.50	Radiative	22
Quartz	<10	160-380	10^{-6}-10^{-8}	1.0	LHS	16
Granite	44-104	290-350	10^{-5}-760	1.13	Probe	25
Silica sand	100-400	170-370	10^{-5}-760	1.60	Diff	24
Silica sand	160-250	300	10^{-2}-760	...	Probe	30
Glass beads	<37-590	150-350	10^{-5}-10^{-6}	1.10-1.80	Radiative	22
Glass beads	10-243	100-500	<10^{-6}	1.20-1.60	DLHS	23
Glass beads	29-470	315	10^{-1}-10^{-4}	1.49-1.51	Comparative	29
Glass beads	177-840	300	10^{-5}-760	1.56-1.66	Probe	30
Glass beads	44-62	200-350	10^{-8}	1.42	LHS	16
Glass beads	29-150	300	10^{-5}-760	1.46	Probe	25
Glass beads	≈590-840	190-270	10^{-5}-760	1.59	Cold plate	28
Glass beads	29-200	300	10^{-5}	1.50	LHS	27
Glass beads	34	140-360	10^{-6}-10^{-10}	1.40	LHS	21
Glass beads	34,715	300	10^{-8}	1.40,1.58	DLHS	38

[a]DLHS--differential line heat source; LHS--line heat source; probe--thermal conductivity probe; diff--measurement of thermal diffusivity; radiative--radiative heat transfer-temperature method; comparative--comparative thermal conductivity method; cold plate--guarded cold plate method.

investigators during 1960–1970. Basalt powders were most
often studied because of their expected similarity to lunar
material; pumice and pumice powders were chosen because of
their vesicular nature and volcanic origin. Chondrites and
tektites were selected because of their possible lunar origin.
Glass beads were examined by many investigators, not because
of their possible similarity to lunar materials, but because
of their uniformity, availability in well-defined sizes, sim-
plicity in terms of packing arrangement, and reproducibility.
The other materials represent a broad range--from amorphous
silicas, through harder glasses and complex minerals, to well-
defined crushed crystalline quartz--and were expected to include almost all potential types of lunar materials.

B. Effect of Gas Pressure on Thermal Conductivity

Figure 1 shows the effect of gas pressure on the thermal
conductivity of pumice powder.[25] At atmospheric pressure, the
thermal conductivity values are within $\pm$ 15% for powders of
three different size ranges. At gas pressures below 10^{-3}

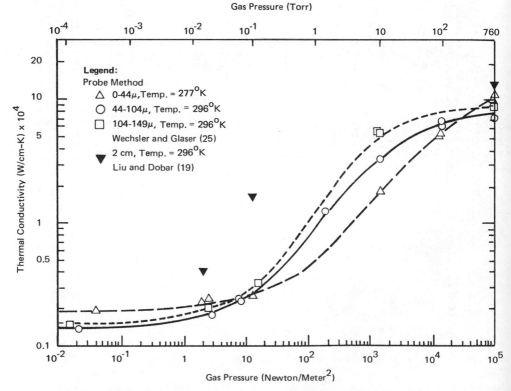

Fig. 1 Thermal conductivity of pumice powder - effect of gas
pressure.

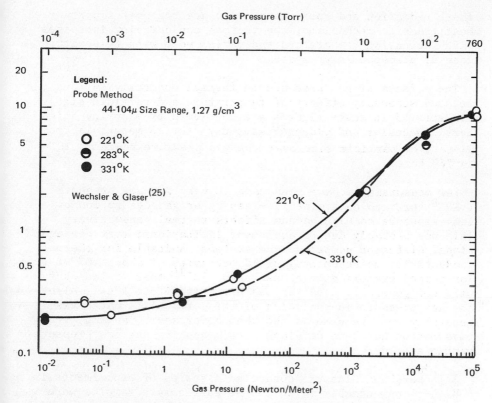

Fig. 2   Thermal conductivity of basalt powder - effect of temperature.

torr, the thermal conductivity values are also within ± 15%. For intermediate gas pressure, there are significant differences among the conductivities for different particle sizes. The greater the particle size, the lower the breakaway pressure. Note that the data obtained by Liu and Dobar,[19] although on a different type of pumice, also show that the greater the particle size, the lower the gas pressure must be to achieve the same reduction in thermal conductivity.

As expected from kinetic theory, gas temperature also affects the shape of the pressure/conductivity curve. Figure 2 shows the results of measurements on basalt powder for three different temperatures. The higher the temperature, the higher the breakaway pressure although this effect is small. Two other important points can be noted from Fig. 2. First, at atmospheric pressure, temperature has little effect on thermal conductivity because gas conduction is the major heat-transfer mechanism and the conductivity of the gas varies only slightly with temperature. Second, at gas pressures below 10^{-2} torr,

where radiation and solid conduction are the most important
heat-transfer mechanisms, the thermal conductivity increases
with increasing temperature and varies more with temperature
than at atmospheric pressure.

The effects of gas pressure on thermal conductivity (and the
related secondary effects of temperature and particle size)
illustrated in these figures are confirmed by many investigators. Wechsler and Glaser[20] gave data for 18 materials from
5μ to 2 cm particle size over the gas pressure range of 10^{-5}
to 760 torr.

Few measurements have been made at very low gas pressures
(<10^{-8} torr) to determine if cleaning or gas desorption by
long exposure to high vacuum affects thermal conductivity.
This was probably due to equipment limitations; most conventional diffusion pumped vacuum systems, suitable for thermal
conductivity measurements, yield pressures of about 10^{-6} torr.
Ion-pumped systems give pressures of 10^{-10} torr or less but
were not generally used for measurements before 1964. Also,
the gas pressure in the voids of the particulate materials is
generally not the same as the chamber pressure, and no effective method has been developed for measuring the gas pressure
within a powder.

Salisbury and Glaser[27] conducted a series of experiments in
a high-vacuum chamber in which the particulate samples were
sieved onto a sample holder in vacuum and baked out under high
vacuum to insure that a low gas pressure was achieved. As
shown in Table 2, very low gas pressure seems to have little
effect on thermal conductivity. These results were confirmed
by Wechsler and Simon,[16] although these investigators did not
sieve the powder in vacuum but baked out the sample at high
temperatures (>200°C) for more than 24 hr. Unpublished measurements by one of the authors (Fountain[21]) indicate only a
slight difference in thermal conductivity when the gas pressure
is decreased from 10^{-6} to less than 10^{-10} torr. These measurements on glass beads were made in an ionization vacuum system
at 10^{-6} torr. The sample was baked at 400°K while still in
vacuum and then allowed to outgas for 14 days after the vacuum
gage had indicated a pressure of less than 10^{-10} torr. Thus,
it was inferred that the high vacuum of the lunar environment
could be simulated for most thermal conductivity measurements
by gas pressures of 10^{-5} torr or less. The effects of solar
and corpuscular radiation, combined with high vacuum, on thermal conductivity of simulated lunar materials were never fully
studied in the laboratory.

Table 2 Effect of very low gas pressure on thermal conductivity

Materials	Particle size, μ	Temperature, °K	Thermal conductivity at		Reference
			10^{-5} to 10^{-6} torr (w/cm-°K)	10^{-9} to 10^{-10} torr (w/cm-°K)	
Olivine	<20	290–294	5.9×10^{-5}	5.9×10^{-5}	27
Chondrite	<20	295	4.0×10^{-5}	4.2×10^{-5}	27
Basalt	104–150	300	1.5×10^{-5}	1.5×10^{-5}	27
Basalt	10–37	297–300	1.79×10^{-5}	1.75×10^{-5}	16
Glass beads	34	360	4.3×10^{-5}	2.7×10^{-5}[a]	21
Glass beads	34	140	1.6×10^{-5}	1.0×10^{-5}[a]	21

[a]Measurements taken at less than 10^{-10} torr.

C. Effect of Materials Type on Thermal Conductivity

Table 3 compares the thermal conductivity values obtained for several different types of materials at gas pressures of less than 10^{-5} torr and at temperatures near room temperature. The two sets of data represent two particle size ranges, between about 30 and 100μ, and <40μ. There is some density variation within this data: the basalt samples used by the different investigators came from different sources; the glass beads came from two manufacturers. For the materials with larger particles, the conductivity values range from 1.0 x 10^{-5} to 4.5 x 10^{-5} w/cm-°K; for the smaller particles, the range is 1.8 x 10^{-5} to 5.9 x 10^{-5} w/cm-°K. No direct relationship between bulk density and thermal conductivity is shown here; this effect is masked by the effects of composition (see later discussion for effects of bulk density on conductivity).

In general, the glassy materials (pumice, glass beads, tektite) have a lower thermal conductivity than the more crystalline materials (olivine, quartz). This lower value is expected if solid conduction contributes significantly to thermal conductivity. Wechsler and Simon[16] interpret their experimental data on several powders to show that the ratio of radiation to conduction heat transfer at 300°K varies from about 0.33 for a small-particle-size quartz powder to about 3.8 for a larger-particle-size pumice powder. These data suggest that most of the variation shown in Table 3 is a result of the varying contributions of conduction and radiation, which depend on composition and particle size. However, the effect of composition on thermal conductivity is not very great compared to the effect of particle size and temperature.

D. Effect of Temperature on Thermal Conductivity

The effects of temperature on the thermal conductivity of particulate materials have been measured by several investigators.[16,22-24,26] The general approach has been to measure thermal conductivity over the range from about 100° to 400°K, fit the data empirically to Eqs. (7a-7c), and evaluate and examine the constants. Practically all data on particulate materials show that thermal conductivity increases with temperature in this range and can best be represented by Eq. (7a), where the constants depend upon material type and particle size. Figure 3 shows representative data for glass microbeads, quartz, and basalt powders. The coefficients representing the solid conduction and radiation contributions to heat transfer are given in Table 4, along with values of these coefficients, for other materials.

Table 3 Thermal conductivity of selected materials at 10^{-5} torr and 300°K^a: effect of composition

Material	Particle size, μ	Density, g/cm^3	Thermal conductivity, w/cm-°K, x 10^5	Reference
Medium-size powders				
Glass beads	53–74	1.38–1.68	1.6	22
Glass beads	44–74	1.42	1.2	16
Glass beads	38–53	1.55	1.2	23
Basalt	10–76	1.60	2.6	24
Basalt	44–104	1.27	2.1	16
Basalt	44–74	1.43	1.0	16
Basalt	37–62	1.30	1.9	26
Crushed quartz	44–74	1.26–1.46	4.5	22
Crushed hornblende	<74	1.11–1.50	3.5	22
Crushed olivine	<74	1.37	4.5	22
Pumice	44–104	0.80	1.4	25
Granite	44–104	1.13	1.7	25
Small-size powders				
Crushed olivine	<20	...	5.9	20
Chondrite	<20	...	4.0	20
Tektite	<20	...	2.3	20
Basalt	<20	...	1.8	20
Pumice	<44	0.88	2.0	20
Quartz	<10	1.00	3.4	16
Glass beads	<37	1.17–1.47	2.6	22

^{a} At pressures of 10^{-5} torr or less and temperatures of 300° ±20°K.

Agreement among various investigators' data depends upon the materials used. Watson's[22] and Merrill's[23] data for glass beads show that the solid conduction term decreases with increasing particle size, and the radiative term decreases with decreasing particle size. These trends can be explained as follows. For a given thickness of powder, the conduction heat flow per unit area should be proportional to the number of contacts per unit area times the conductance per contact path. For an orderly array of spheres, the number of contacts per unit area is approximately inversely proportional to the square of the particle diameter $(1/d^2)$. The conductance per contact path is inversely proportional to the number of contacts which is, in turn, inversely proportional to the diameter; therefore, the conductance per contact path is proportional to $(1/d)^{-1}$. Thus the heat flow by conduction is proportional to $(d^2)^{-1}$ x $(1/d)^{-1} = d - 1$. Watson's data indicate that the solid conduction term $B \simeq 3000/d$, where B is in ergs/cm-sec-°K and d is in microns. The smaller the particle size, the greater the

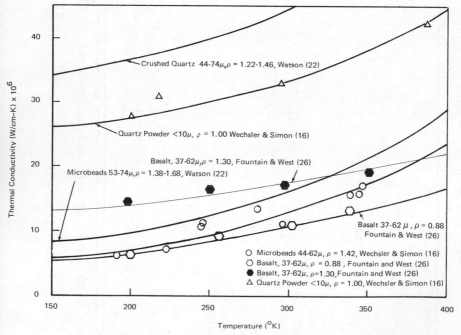

Fig. 3 Effective thermal conductivity of glass beads, basalt, and quartz powder.

Table 4 Conduction and radiation contributions to thermal conductivity

Material	Particle size, μ	Density, g/cm³	Solid conduction (B), contribution w/cm-°K, x 10⁶	Radiation contribution (A), w/cm-°K⁴, x 10¹³	Ratio of radiation/conduction 200°K	300°K	400°K	Reference
Basalt	10-37	1.36	$21-1.6 \times 10^3/T^a$	0.88	0.05	0.15	0.34	16
	44-74	1.43	6.10	2.10	0.28	0.94	2.20	16
	37-62	0.79	5.10	1.70	0.26	0.90	2.10	26
	37-62	0.88	6.50	1.70	0.21	0.71	1.70	26
	37-62	0.98	6.20	1.80	0.23	0.78	1.90	26
	37-62	1.10	8.90	1.90	0.17	0.58	1.40	26
	37-62	1.30	12.40	2.40	0.15	0.52	1.20	26
	37-62	1.50	16.20	3.40	0.17	0.57	1.30	26
Quartz	<10	1.00	25.00	3.00	0.10	0.33	0.77	16
	44-74	1.30	33.00	4.20	0.10	0.34	0.81	22
Glass beads	<37	1.2-1.5	9.50	6.30	0.53	1.80	4.20	22
	53-74	1.4-1.7	7.00	3.40	0.39	1.30	3.10	22
	88-125	1.60	3.20	8.50	2.10	7.16	17.00	22
	250-350	1.5-1.6	0.95	13.00	10.90	37.00	87.70	22
	590-840	1.6-1.8	(-0.66)ᵇ	26.00	...	...	...	22
	44-62	1.40	4.70	3.00	0.51	1.70	4.10	16
	10-20	1.50	4.70	2.80	0.48	1.60	3.80	23
	38-53	1.50	4.50	3.30	0.59	1.90	4.70	23
	125-243	1.30	0.07ᶜ	5.40	61.70	208	493	23
Pumice	10-37	...	5.10	3.10	0.49	1.70	3.90	16
	44-74	...	2.50	3.60	1.10	3.80	9.10	16
Olivine	<74	1.37	10.80	1.30	0.096	0.33	0.77	22

ᵃData were best fit by Eq. (7c).

ᵇA negative value of the conduction term was obtained; this is not possible but indicates that conduction was small.

ᶜSeveral values ranging from -0.09 to +0.15 were obtained for three similar samples; the radiation term varied from 3.1 to 5.4 for these samples.

number of scattering and reflective surfaces per unit length
in the powder; thus, the radiation coefficient should decrease
with decreasing particle size. Wechsler and Simon's[16] data
show similar results for basalt and pumice powder.

Comparison of the data for glassy and crystalline materials
shows that the crystalline materials (quartz or olivine, for
example) have solid conduction contributions higher than glass
beads or pumice powder. The radiation contribution, in addition to being a function of particle size, depends upon the
optical constants of the material. Wechsler and Simon[16] show
that the radiative term can be calculated (to within about
100%) if the optical properties are known, and that experimental data for the radiative contribution agree reasonably well
with an empirical relation of Godbee and Ziegler.[31] Merrill[23]
has compared the measured radiative contributions with several
empirical expressions based on porosity and particle size with
good results. Fountain and West's[26] data on basalt show that
the solid conductive contributions increase with bulk density
as expected (greater number of contacts), but, for unknown
reasons, the radiative contribution also increases. This apparent increase may actually have been a temperature-dependent
increase in the solid conduction term [Eq. (7b)] which does
not appear in Eq. (7a), from which these coefficients were obtained.

A most important conclusion to be drawn from Table 4 is that,
over the range of lunar temperatures, the ratio of radiative/
conductive heat transfer changes considerably for most powders.
Therefore, solid conduction may be most significant at lunar
night and radiation most important during the lunar day.

E. Effect of Density on Thermal Conductivity

Perhaps the most lucid demonstration of the effect of bulk
density on thermal conductivity was made by Fountain and
West,[26] who studied an Oregon basalt sample of 37-62 particle
size. The sample was baked out in vacuum at 525°K and sifted
onto a line heat source apparatus. Bulk densities over the
range of 0.79 to 1.50 g/cm^3 were obtained by adding more powder to the sample, and vibrating the sample container and powder to compact the powder. Conductivity values for two different bulk densities were shown in Fig. 3 as a function of
temperature. Figure 4, taken from the data of Fountain and
West,[26] shows that the thermal conductivity increases almost
in direct proportion to the bulk density above about 1 g/cm^3.
The original data show an anomaly in the region from 0.8 to
1.0 g/cm^3 caused by the sample of 0.98 g/cm^3 density, the
thermal conductivity of which was lower than that of a sample

of 0.88 g/cm³ density. These data have been corrected for a
slight numerical error in curve fitting. Re-examination of
the data suggests that the scatter in the experimental data
could account for this result, and thus we have drawn a smooth
curve for the data below 1.0 g/cm³. It may, however, be possible that this behavior is real and that it can be attributed
to the interplay between the radiative and conductive contributors to heat transfer.

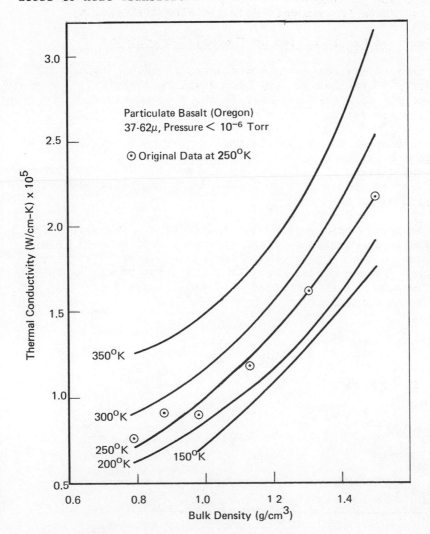

Fig. 4   Thermal conductivity of basalt powder - effect of
         density. Note: curves are smoothed curves. Redrawn
         from data by Fountain and West[26].

A minimum in the conductivity/density relationship was noted by Everest[7] in studies of thermal conductivity of evacuated expanded perlite and colloidal silica insulations, even though the bulk density range was much lower. Thus, one can conclude that at bulk densities above about 1.0 g/cm³ the thermal conductivity of most silicates should increase proportionately to density; at lower bulk densities, the conductivity depends on the structural arrangement of the powder and on the composition, and these in turn determine the effect of density.

F. Effect of Particle Size on Thermal Conductivity

The interaction between gas pressure, thermal conductivity, and particle size has already been discussed briefly. The effects of particle size can best be shown by considering the data on glass beads.

Figure 5 shows the data obtained by several investigators at atmospheric pressure and at gas pressures below 10^{-2} torr, where the effects of gas pressure are small. Although there appears to be wide variation in the data, several definitive conclusions can be drawn:

1) At atmospheric pressure, the data of all investigators are relatively consistent with respect to the absolute value of thermal conductivity. This consistency is attributed to the large gas conduction contribution at atmospheric pressure. There is only a small increase of thermal conductivity with an increase in particle size.

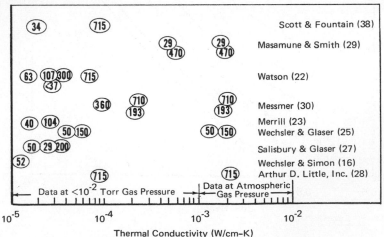

Fig. 5   Effect of particle size on thermal conductivity. Note: numbers refer to particle diameter in microns; all data reported at approximately 300°K.

2) Under high vacuum conditions, thermal conductivity varies considerably with particle size. The results of Watson,[22] Wechsler and Glaser,[25] Salisbury and Glaser,[27] Arthur D. Little, Inc.,[28] Merrill,[23] and Scott and Fountain[38] are consistent in both the absolute value of thermal conductivity and the increase of thermal conductivity with increased particle size. The thermal conductivity of beads with very small particle size (Watson,[22] Salisbury and Glaser,[27] and Scott and Fountain[38]) appears to be higher than that of beads with an intermediate particle size. These trends can be explained by the different values of the conductive and radiative contributions to thermal conductivity shown previously in Table 4. Conduction decreases and radiation increases with increasing particle size; at 300°K, radiation is the more important mechanism.

3) The source (or composition) of materials has no consistent effect. Masamune and Smith,[29] Wechsler and Glaser,[25] and Salisbury and Glaser[27] used 3M glass beads; others used Microbeads, Inc., material. The comparative method of Masamune and Smith[29] gives higher thermal conductivity values and less effect of particle size on thermal conductivity than any other method. The thermal conductivity probe methods used by Messmer[30] and Wechsler and Glaser[25] give higher thermal conductivity values than the line heat source or guarded flat-plate methods. These data suggest that the probe and comparative methods may have lower limits of applicability or may yield inaccurate data at low conductivity values (low gas pressures). Work by one of the authors (Fountain[21]) showed that measurements on a single, well-evacuated, low-conductivity sample by the line heat source method yielded significantly lower conductivity values than measurements by the probe method. The high value of conductivity at 10^{-2} torr which Messmer[30] obtained for beads with a 193-μ diam does not agree with the other observed trends.

4) Beads that have been thoroughly cleaned and outgassed at low pressures (Watson[22] and Wechsler and Simon[16]) seem to have lower thermal conductivity values than those for which the preparation method is not specified. This difference may be a coincidence, or it may be the result of removing conducting films on the beads by cleaning or prolonged outgassing.

5) The bulk densities of the glass-bead samples used in these measurements varied from about 1.4 to 1.7 g/cm^3, with the exception of the low density, 1.3 g/cm^3, of the sample of small particle size used by Watson.[22] The trends of particle

size shown in the figure seem to be independent of density
variations. The low density of Watson's sample could lead to
a higher radiation contribution and therefore the higher conductivity.

G. Effect of Mechanical Load on Thermal Conductivity

Watson[22] suggested that the solid conduction contribution to
thermal conductivity could be calculated from knowledge of the
properties of the bulk material--its density, conductivity,
thickness, particle size, and mechanical properties--based upon
a Hertzian loading theory. However, his calculations were not
confirmed by his experiments with glass beads, the calculations
did not show as large an effect of particle size as demonstrated experimentally. Wechsler and Simon's[16] results show
that the values of solid conduction contribution for glass
beads, pumice, and basalt can be calculated to within a factor
of 2.5 by Watson's approach.

Messmer[30] also investigated the effect of compressive loading
on the thermal conductivity of glass beads. Evacuated glass
beads of 710- and 193-μ diam were subjected to external pressures of 10^5 to 7×10^6 N/m^2 at 300°K. The thermal conductivity of both samples increased with the 1/4 power of the applied
external pressure. This relationship differs slightly from the
1/3 power of the external pressure obtained from Hertzian loading analysis. Data obtained at Arthur D. Little, Inc.,[28] using
710-μ-diam evacuated glass beads at 225°K in a guarded cold-
plate apparatus, suggested that the thermal conductivity increases with about the 1/10 power of the load over the pressure
range from 5×10^3 to 3×10^4 N/m^2. Data by one of the authors
(Fountain[21]) on glass beads also show an increase in thermal
conductivity with mechanical load of the same magnitude as observed by Messmer (see Chapter 2.c).

The radiation component of thermal conductivity should be
almost invariant with loading, whereas the conduction component
should increase with the 1/3 power of the load. Therefore, the
dependence of thermal conductivity on external load varies in
accordance with the ratio of solid conduction to radiation.

Glaser and Wechsler[32] investigated the effects of reduced
gravity on the thermal properties of particulate materials by
measuring heat transfer in evacuated powders during aircraft
flights in 1/6-g and 0-g trajectories. The results indicated
a decrease in thermal conduction of up to 30% in aluminum oxide
and expanded perlite particles of 180- to 400-μ size. On the
basis of these results, it can be concluded that the thermal
conductivity of possible lunar materials will depend more on

material density, composition, particle size, and temperature
than on the 1/6-g lunar environment.

IV.   Specific Heat of Simulated Lunar Materials

The specific heat of particulate materials is a function of
composition, phase, crystalline arrangement, and temperature.
As mentioned earlier, the variation of specific heat over the
expected range of lunar temperatures and compositions is much
less than the variation in thermal conductivity; therefore,
only a few experimental measurements of specific heat during
the period of 1960–1970 were directed at a better understanding
of lunar surface materials.

The specific heat of most minerals has been fairly well established over the temperature range from room temperature to
several hundred degrees Centigrade. Data are usually presented
as a function of temperature as follows:

$$c_p = a + bT - (C/T^2) \qquad (8)$$

Values of specific heat for more than 100 minerals are given by
Birch.[33]

The specific heat of rocks has been studied less extensively;
emphasis has been placed upon determining the mineralogical
composition and computing their specific heat from that of con-
stituent materials. At room temperature, the specific heats of
most rocks vary from about 0.14 to 0.25 cal/g-K. Only those
rocks with a high water content (serpentine, pumice) have
values above about 0.21 cal/g-K.

The specific heat of minerals increases with increasing temperature because of increased vibrations of the atoms about
their lattice position. The specific heat of glasses also increases with temperature because of thermal motions. Figure 6
shows representative values of specific heat of several glasses
and rocks. Note that, over the lunar temperature range, the
specific heat of these materials varies from about 0.1 to 0.24
cal/g-K, the glassy materials having the lower values and the
more crystalline materials having the higher values.

Specific heat data for returned Apollo 11 samples have already been reported by Robie et al[39] (see Chapter 3.b).

V. Thermal Diffusivity and Thermal
Parameter of Simulated Lunar Materials

With the exception of the results reported by Bernett et al.,[24] thermal diffusivity ($k/\rho c_p$) and thermal parameter $(k\rho c_p)^{-1/2}$ have rarely been measured using simulated lunar materials. Instead, they have been calculated from measured thermal conductivity and measured or estimated specific heats and densities.

Table 5 shows typical values of thermal diffusivity and thermal parameter for a variety of granular materials. The values are arranged to illustrate the effects of selected variables on these properties. The variation in thermal parameter values with material type or composition encompasses the range of values estimated from observational data. The materials of smaller particle size generally have lower values of the thermal parameter, but only because the thermal conductivities and densities chosen were lower than for the samples of large particle size. The more crystalline materials tend to have lower values of the thermal parameter than the more amorphous materials because of their generally higher thermal conductivities. The effect of material type on thermal diffusivity shows the general trend of greater diffusivity for the more crystalline materials.

The effect of density on thermal parameter is shown for basalt powder; because the thermal conductivity increases with increasing density, the thermal parameter decreases signifi-

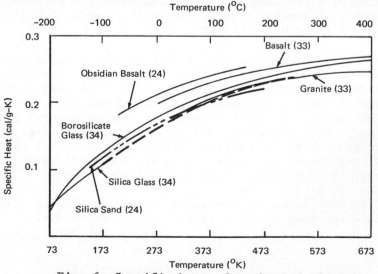

Fig. 6 Specific heat of rocks and glasses.

Table 5 Thermal diffusivity and thermal parameter of particulate materials

Material	Particle size, μ	Density, g/cm³	Temperature, °K	Thermal diffusivity, cm²/sec × 10⁵	Thermal parameter, cm²·°K·sec^(1/2)/cal	Reference
Effect of composition						
Basalt	44–74	1.40	300	0.81	1190	16
Glass beads	44–74	1.40	300	1.40	1290	16
Crushed quartz	44–74	1.36	300	4.00	590	22
Olivine	<74	1.37	300	3.80	580	22
Hornblende	<74	1.30	300	3.10	680	22
Pumice	44–104	0.80	300	2.00	1350	25
Granite	44–104	1.13	300	2.10	1120	25
Olivine	<20	...	300	...	510	27
Chondrite	<20	...	300	...	720	27
Basalt	<20	...	300	...	950	27
Pumice	<44	0.88	300	2.60	950	27
Quartz	<10	1.00	300	3.90	620	16
Glass beads	<37	1.32	300	3.10	900	22
Effect of density						
Basalt	37–62	0.79	300	1.30	1670	26
	37–62	0.88	300	1.40	1430	26
	37–62	1.13	300	1.40	1140	26
	37–62	1.30	300	1.60	910	26
	37–62	1.50	300	1.90	710	26
Effect of particle size						
Glass beads	<37	1.32	300	3.10	900	22
	53–74	1.53	300	1.70	1070	22
	88–125	1.60	300	2.60	820	22
	250–350	1.55	300	4.40	700	22
	590–840	1.71	300	6.60	480	22
Pumice	<44	0.88	300	3.30	1000	20
	44–104	0.80	300	1.80	890	20
	104–149	0.84	300	2.90	590	20
Effect of temperature						
Basalt	37–62	1.30	200	1.50	1040	26
	37–62	1.30	300	1.70	890	26
	37–62	1.30	400	2.30	730	26
Glass beads	44–62	1.40	200	0.80	2040	16
	44–62	1.40	300	1.40	1290	16
	44–62	1.40	400	1.90	770	16
Quartz powder	44–74	1.36	200	4.90	2010	16
	44–74	1.36	300	4.00	1410	16

cantly. Less effect is noted for thermal diffusivity, because it represents the ratio of conductivity to density.

Materials of larger particle size generally have a lower thermal parameter and a higher thermal diffusivity, because the larger powders have a greater thermal conductivity. (The increase in thermal conductivity is greater than the increase in corresponding density for larger-particle-size materials.)

Thermal parameter decreases and thermal diffusivity increases with increasing temperature because of the increase in thermal conductivity with temperature. These results indicate that analysis of infrared temperatures of the lunar surface and their interpretation in terms of lunar materials must take into account the change in materials properties with temperature, particularly the temperature dependence of thermal conductivity.

Early interpretations of thermal parameter values obtained by Jaeger,[3] and Jaeger and Harper[2] (1000-1300) indicate a material of low density and low conductivity, e.g., low-density basalt or pumice. Later studies by Russian investigators (Troitskii,[35] Krotikov, and Troitskii,[36] and Krotikov and Shchuko[37]) giving values of 350-500 indicate that a more dense, crystalline material of larger particle size would be more representative of the lunar surface. For models in which constant thermal properties are assumed, the value of γ which best fits nighttime cooling is 800; a value of γ of 1300 best fits eclipse data (see Chapter 1.a). These values indicate the presence of a range of surface materials several meters deep of intermediate density and thermal conductivity, perhaps of density 1.2 to 1.5 g/cm^3 and thermal conductivity of 1.0 to 4.0 x 10^{-5} w/cm-°K.

References

[1] Wesselink, A. J., "Heat Conductivity and the Nature of the Lunar Surface Material," Bulletin of the Astronautical Institute of the Netherlands, Vol. 10, 1948, p. 351.

[2] Jaeger, J. C. and Harper, A. F. A., "Nature of the Surface of the Moon," Nature, Vol. 166, No. 4233, 1950, p. 1026.

[3] Jaeger, J. C., "The Surface Temperature of the Moon," Australian Journal of Physics, Vol. 6, 1950, p. 10.

[4] Linsky, J. L., "Models of the Lunar Surface Including Temperature Dependent Thermal Properties," Scientific Rept. 8, 1966, Harvard College Observatory.

[5] Halajian, J. D., Reichman, J., and Karafiath, L. L., "Correlation of Mechanical and Thermal Properties of Extraterrestrial Materials," report prepared for NASA/Marshall Space Flight Center by Grumman Aircraft Engineering Corp. under Contract NAS8-20084, Jan. 1967.

[6] Smoluchowski, M., "Sur la Conductibilite Calorifique des Corps Pulverises," Bull. Int. de L'Academic des Sciences de Cracovie, Vol. A, 1910, p. 129.

[7] Everest, A., Glaser, P. E., and Wechsler, A. E., "On the Thermal Conductivity of Powder Insulations," XI International Congress on Refrigeration, Aug. 1963.

[8] Schotte, W., "Thermal Conductivity of Packed Beds," American Institute of Chemical Engineers Journal, Vol. 6, No. 1, 1960, pp. 63-67.

[9] Deissler, R. G. and Eian, C. S., Res. Memo RME 52Co5, June 24, 1952, NACA.

[10] Little, R. C., Carpenter, F. G., and Deitz, V. R., "Heat Transfer in Intensively Outgassed Powders," Journal of Chemical Physics, Vol. 37, No. 8, 1962, pp. 1896-1898.

[11] Drabble, J. R. and Goldsmid, H. J., Thermal Conduction in Semiconductors, Pergamon Press, New York, 1961.

[12] Holm, R., "Thermal Contacts," report to NASA/George C. Marshall Space Flight Center, Huntsville, Ala., Oct. 1962.

[13] Strong, H. M., Bundy, F. P., and Bovenkerk, H. P., "Flat Panel Vacuum Thermal Insulation," Journal of Applied Physics, Vol. 31, No. 1, 1960, pp. 39-50.

[14] Russell, H. W., "Principles of Heat Flow in Porous Insulators," Journal of the American Ceramic Society, Vol. 18, 1935, pp. 1-5.

[15] Fremlin, J. H., "Subsurface Temperatures on the Moon," Nature, Vol. 183, No. 4656, 1959, p. 239.

[16] Wechsler, A. E. and Simon, I., "Thermal Conductivity and Dielectric Constant of Silicate Materials," final report under NASA Contract NAS8-20076, 1966.

[17]Chen, J. C. and Churchill, S. W., "Radiant Heat Transfer in Packed Beds," American Institute of Chemical Engineers Journal, Vol. 9, No. 1, 1963, pp. 35-41.

[18]Wechsler, A. E., Glaser, P. E., and Allen, R. V., "Thermal Conductivity of Non-Metallic Materials," summary report prepared by Arthur D. Little, Inc., for George C. Marshall Space Flight Center under Contract NAS8-1567, 1963.

[19]Liu, N. C. and Dobar, W. I., "The Nature of the Lunar Surface: The Thermal Conductivity of Dust and Pumice," The Lunar Surface Layer, edited by J. W. Salisbury and P. E. Glaser, Academic Press, New York, 1964.

[20]Wechsler, A. E. and Glaser, P. E., "Pressure Effects on Postulated Lunar Materials," Icarus, Vol. 4, No. 4, Sept. 1965, pp. 335-352.

[21]Fountain, J. A., unpublished results, 1971.

[22]Watson, K., "Thermal Conductivity of Selected Silicate Powders in Vacuum from 150°-350°K," Thesis, 1964, California Institute of Technology.

[23]Merrill, R. B., "Thermal Conduction Through an Evacuated Idealized Powder Over the Temperature Range of 100° to 500°K," TN D-5063, March 1969, NASA.

[24]Bernett, E. C., Wood, H. L., Jaffe, L. D., and Martens, H. E., "Thermal Properties of a Simulated Lunar Material in Air and in Vacuum," AIAA Journal, Vol. 1, No. 6, 1963, pp. 1402-1407.

[25]Wechsler, A. E. and Glaser, P. E., "Thermal Conductivity of Non-Metallic Materials," Summary Report, Contract NAS8-1567, 1964, Arthur D. Little, Inc.

[26]Fountain, J. A. and West, E. A., "Thermal Conductivity of Particulate Basalt as a Function of Density in Simulated Lunar and Martian Environments," Journal of Geophysical Research, Vol. 75, No. 20, July 10, 1970, p. 4063.

[27]Salisbury, J. W. and Glaser, P. E., eds., Studies of the Characteristics of Probable Lunar Surface Materials, AFCRL-64-970, Jan. 1964, Air Force Cambridge Research Laboratories.

[28]Arthur D. Little, Inc., unpublished work conducted for the Bendix Corporation under subcontract SC 0242 for the National Aeronautics and Space Administration under NAS9-5829, 1969.

[29] Masamune, S. and Smith, J. M., "Thermal Conductivity of Beds of Spherical Particles," Industrial and Engineering Chemistry, Vol. 2, No. 2, 1963, pp. 136-142.

[30] Messmer, J., personal communciation, 1966, Gulf Research and Development Corp.

[31] Godbee, H. W. and Ziegler, W. T., "Thermal Conductivity of MgO, Al_2O_3 and ZrO_2 Powders to 850°C," Journal of Applied Physics, Vol. 37, No. 1, 1966, pp. 40-55.

[32] Wechsler, A. E. and Glaser, P. E., "Investigation of the Effects of Reduced Gravity on the Thermal Properties of Insulation Materials," Summary Report under NASA Contract No. NAS8-5413, Jan. 1965.

[33] Birch, F., ed., "Handbook of Physical Constants," Special Paper 36, 1942, Geological Society of America.

[34] Goldsmith, A., Hirschhorn, H. J., and Waterman, T. E., "Thermophysical Properties of Solid Materials," Ceramics, TR 58-476, Vol. III, Nov. 1960, Wright Air Development Center.

[35] Troitskii, V. S., "A New Possibility of the Determination of the Density of the Surface Rocks of the Moon," Radiophysics, Vol. 5, No. 5, 1962a, p. 885.

[36] Krotikov, V. D. and Troitskii, V. S., "Radio Emission and Nature of the Moon," Soviet Phys.-Uspekhi, Vol. 6, No. 6, 1963a, pp. 841-871.

[37] Krotikov, V. D. and Shchuko, O. B., "The Heat Balance of the Lunar Surface Layer During Lunation," Soviet Astronomy, AJ, Vol. 7, No. 2, 1963, p. 228.

[38] Scott, R. W. and Fountain, J. A., "A Comparison of Two Transient Methods of Measuring Thermal Conductivity of Particulate Samples," TMX-64559, Sept. 1970, NASA.

[39] Robie, R. A., Hemingway, B. S., and Wilson, W. H., "Specific Heats of Lunar Surface Materials from 90 to 350 Degrees Kelvin," Science, Vol. 167, No. 3918, Jan. 30, 1970, pp. 749-750.

THERMAL PROPERTY MEASUREMENTS ON LUNAR MATERIAL
RETURNED BY APOLLO 11 AND 12 MISSIONS

Ki-iti Horai* and Gene Simmons[/]

Lunar Science Institute, Houston, Texas
Massachusetts Institute of Technology, Cambridge, Mass.

I. Introduction

On July 20, 1969, the Apollo 11 lunar module landed in the
southwestern part of Mare Tranquillitatis at $0.67°N$ and $23.49°$
E. A total of 21.5 kg of lunar material was collected during
the extravehicle activity and was returned to the earth.

According to the mode of collection, the Apollo 11 lunar
samples are divided into three categories.[1] The contingency
sample was collected at the earliest stage of extravehicle
activity to secure a certain amount of lunar materials to be
returned to the earth. The bulk sample is the lunar material
collected in one of the two sample containers for the purpose
of furnishing a large quantity of lunar material. The documented sample is the lunar material collected in another sample
container with the full documentation of the rock samples with
regard to the location, orientation, and local geological setting. The two drive core tubes, gas analysis sample, and lunar
environmental sample also are included in this category.

Each of the forementioned Apollo 11 lunar materials consists
of rocks and soil. The Lunar Sample Preliminary Examination
Team,[1] which undertook the preliminary examination of the lunar
material, made a preliminary classification of the samples.
According to this classification, type A refers to the fine-
grained vesicular igneous rocks, type B to the medium-grained
vuggy igneous rocks, type C to the breccia, and type D to the
fines, which are, by definition, the fragments less than 1 cm
in diameter.

Two drive tube core samples will deserve a somewhat detailed
description.[2,3] They are the samples of lunar fines obtained

---

*Visiting Scientist (Geophysicist).
[/]Professor of Geophysics, Department of Earth and Planetary
Sciences.

by driving the aluminum tube, 31.75 cm in length and 1.95 cm
in diameter, into the surface of lunar regolith. Core no. 1
(sample 10005-0), 10 cm long, contains 22.39 g of material.
Core no. 2 (sample 10004-0), 13.5 cm long, is 26.73_3g. The
bulk densities are 1.70 ± 0.04 and 1.58 ± 0.04 g/cm^3, respectively. The core material is loose, weakly cohesive, and consists of single grains except for minor aggregates of glass.
Texturally, the core consists of a silty fine sand with the
average grain size about 0.11 mm. Admixed with the sandy
matrix are the fragments of rock, glass spherules, and aggregates of glass with the maximum size of 3 mm or more. The
largest glass aggregate contained in core no. 2 is 1.2 cm in
diameter. The lunar fines are characterized by high glass
content, about 50%.

The lunar samples were distributed to the selected investigators for the detailed study of petrology, mineralogy, chemistry, and physical properties. As a result of this, more detailed and satisfactory classification of the individual samples became possible. Some of the samples were used for the
study of thermophysical properties. Thermal diffusivity,
thermal conductivity, and specific heat have been measured and
their significance and implication discussed. Table 1 summarizes the present status of thermophysical property investigation. A brief description of the samples used for the study
will be given in the Appendix.

The Apollo 12 lunar module landed on the northwest rim of the
Surveyor crater at 23.43°W and 2.45°S on Nov. 19, 1969. A
total of 34.3 kg of lunar materials, 1.9 kg of contingency sample, 14.8 kg of selected sample (same category as Apollo 11
bulk sample), 11.1 kg of documented sample, and 6.5 kg of tote
bag sample was collected and returned to the earth.[5] They contain 4.5 kg of fines (materials less than 1 cm in size), less
than 1.0 kg of chips (materials between 1 and 4 cm in size),
29.0 kg of rocks (materials larger than 4 cm in size), two core
tubes (19 and 40 cm in length), lunar environment sample, and
gas analysis sample.

Reflecting the different geological environments of the landing sites, the Apollo 12 materials are contrasted with the
Apollo 11 materials in various ways. For example, the Apollo
12 rock samples are predominantly crystalline as opposed to the
Apollo 11 rocks, about half of which are microbreccia. The
modal mineralogy and the primary texture of Apollo 12 crystalline rocks show much wider varieties than those of Apollo 11
crystalline rocks. The lunar regolith, which is 3 to 6 m in
thickness at the Apollo 11 landing site, is about one-half as
thick at the Apollo 12 landing site.

Table 1 Present status of thermophysical property measurement for Apollo 11 lunar sample

Thermophysical property	Type		
	A	C	D
Thermal diffusivity κ	Measured [6,7] 10020 (171°~415°K, 1 atm) 10057 (149°~436°K, 1 atm)	Measured [6,7] 10046 (162°~433°K, 1 atm) 10065 (178°~414°K, 1 atm)	
Specific heat C_p	Measured [10,11] 10057 (96°~348°K, 1 atm) Measured [15] 10017 (2.3°~5.0°K, 1 atm)	Measured [15] 10046 (3.1°~4.1°K, 1 atm)	Measured [10,11] 10084 (95°~343°K, 1 atm)
Thermal conductivity k	Estimated (Fig. 3) (120°-480°K, 1 atm)	Measured [15] 10046 (4°K, 1 atm) Estimated (Fig. 3) (120°-480°K, 1 atm)	Measured [18,19] 10084 (203°~405°K, $10^{-2}\sim10^{-7}$ torr)
Thermal parameter γ	Estimated (Fig. 4) (120°-140°K, 1 atm)	Estimated (Fig. 4) (120°-480°K, 1 atm)	Estimated (Table 7) (200°~400°K, 10^{-3} torr)

   Study of thermophysical properties, as well as other physical
properties of Apollo 12 samples, is underway. It is expected
that the Apollo 12 materials with a variety of petrology,
mineralogy, and chemistry should exhibit thermophysical properties that are distinctive from Apollo 11 materials.

           II. Thermophysical Properties of Lunar Material

A. Thermal Diffusivity

   Thermal diffusivity of four Apollo 11 lunar samples, 10020,
10046, 10057, and 10065, was measured over the temperature
range 150° to 440°K by the modified Angstrom method.[6,7] Figure
1 summarizes the result of measurements. Of the four samples
used for the study, samples 10020 and 10057 are type A with
densities 2.99 and 2.88 g/cm^3. Samples 10046 and 10056 are
type C with densities 2.21 and 2.36 g/cm^3. The temperature
variation of thermal diffusivity is almost identical for samples 10020 and 10057 and for samples 10046 and 10065 but is
quite distinctive for each of these two groups.

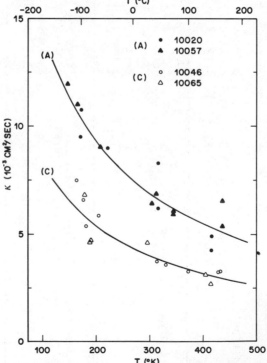

Fig. 1 Thermal diffusivity vs temperature for Apollo 11 lunar
        materials. Closed circles, sample 10020; closed tri-
        angles, 10057 (type A). Open circles, 10046; open
        triangles, 20065 (type C).[7]

According to the experimental study by Kanamori et al.[24], thermal diffusivity of rock-forming mineral is an inverse function of temperature. Coefficients of a linear relationship were determined by least-squares from the data. For type A samples,

$$\varkappa^{-1} = (0.314 \pm 0.159) \times 10^2 + (0.378 \pm 0.051)T \qquad (1)$$

and, for type C samples,

$$\varkappa^{-1} = (0.545 \pm 0.207) \times 10^2 + (0.648 \pm 0.068)T \qquad (2)$$

where thermal diffusivity $\varkappa$ is in square centimeters per second and temperature T is in degrees Kelvin. Values of thermal diffusivity smoothed by relationships (1) and (2) are listed, at 20°K intervals, in Table 2.

Table 2 Thermal diffusivities (smoothed values) for Apollo 11 lunar samples 10020 and 10057 (type A) and 10046 and 10065) (Type C)[6,7]

Temperature, ($^{\circ}$K)	Thermal diffusivity, (10^{-3} cm^2/sec)	
	Type A (10020, 10057)	Type C (10046, 10065)
140	11.86	6.89
160	10.88	6.32
180	10.06	5.84
200	9.35	5.43
220	8.73	5.08
240	8.12	4.76
260	7.71	4.49
280	7.29	4.24
300	6.91	4.02
320	6.56	3.82
340	6.25	3.64
360	5.97	3.48
380	5.71	3.33
400	5.48	3.19
420	5.26	3.06
440	5.06	2.94

The thermal diffusivity of type C samples is lower and less temperature-dependent than type A samples. The difference may be explained in terms of texture and mineral composition of the samples. Porosity of sample 10057 (type A), obtained from the

point count of cavities on the surfaces of the specimen, is
0.174.[8,9] Densities of samples 10046 and 10065 smaller than
those of samples 10020 and 10057 suggest that type C materials
are even more porous than type A. General examination of samples shows that closely spaced microfractures predominate in
type C samples.[1] Type C samples also show various degrees of
vitrification. Table 3 is the result of modal analysis of samples 10020 (type A) and 10046 (type C).[6,7] The presence of
glass, as well as more abundant pores and microcracks in type
C samples is probably the cause of the lower thermal diffusivity and small temperature dependency.

Table 3 Mineral composition of Apollo 11 lunar samples [6,7]

Mineral	Volume fraction, %
a) 10020[a]	
Pyroxene	45.4
Plagioclase	24.6
Olivine	3.9
Ilmenite	22.7
Troilite	0.9
Other	0.7
Void	1.8
b) 10046[b]	
Pyroxene	16.8
Opaques	8.6
Plagioclase	4.7
Unidentified	3.5
Glass	2.9
Matrix (40μ)[c]	63.5

[a]Measured in transmitted (598 points counted) and reflected
(322 points counted) light. Ilmenite contains some
ferropseudobrookite.

[b]Measured in transmitted light (1571 points counted).

[c]The approximate composition of the matrix is 30% glass,
55% pyroxene, and 15% plagioclase.

B. Specific Heat

Specific heat, or heat capacity, of Apollo 11 samples 10057
(type A) and 10084 (lunar soil, type D) was measured on the
temperature range between 95° and 348°K.[10,11] An adiabatic
calorimeter, especially designed and constructed for the study
of lunar specimens,[11,12] was used for the measurements. Figure

2a shows the data for sample 10057. The data for sample 10084
are similar (Fig. 2b). The data are smoothed by fitting or-
thogonal polynomials. The smoothed values of specific heat are
listed in Table 4. It is noteworthy that samples 10057 and
10084 exhibit almost identical specific heats regardless of the
different mineralogy of these specimens.

Independent of the forementioned work, the specific heat of
sample 10064 (type C), 0.20 ± 0.02 cal/g-°K, was reported
by Bastin et al.[13,14] The method of measurement and the temperature at which the measurement was made are not clear, however.

Recently, the specific heat of samples 10017 (type A) and
10046 (type C) was measured at liquid helium temperatures by
Morrison and Norton.[15] The calorimeter used for the measurements consists of a light circular copper tray about 2 cm in
diameter attached to a nitrogen-filled germanium thermometer
and an electric heater. The specific heat of sample 10017
increases almost monotonically from 4.62×10^{-3} cal/g-°K to
6.7×10^{-3} cal/g-°K on the temperature range between 2.344° and
4.97°K. The specific heat of sample 10046, measured on the

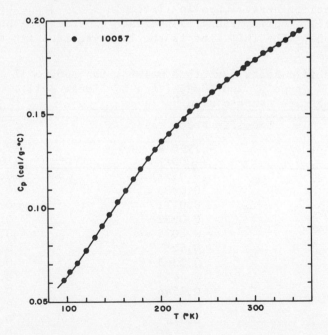

Fig. 2a   Specific heat of Apollo 11 sample 10057 (type A).
          Solid circles indicate experimental observations. The
          full line is the least-squares fit to the data.[11]

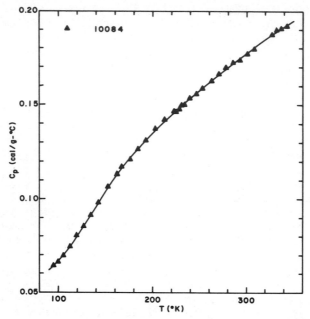

Fig. 2b   Specific heat of Apollo 11 sample 10084 (type D, lunar
          soil).   Solid triangles indicate experimental observa-
          tions.   The full line is the least-squares fit to the
          data.[11]

Table 4   Specific heats (smoothed values) for Apollo 11 lunar
samples 10057 (type A) and 10084 (type D: lunar soil).   Values
in parentheses are extrapolated[10,11]

Temperature, $^{\circ}$K	Specific heat, cal/g-$^{\circ}$K	
	Type A (10057)	Type D (10084)
90	(0.0571)	(0.0615)
100	0.0633	0.0665
120	0.0771	0.0802
140	0.0922	0.0955
160	0.1075	0.1108
180	0.1217	0.1235
200	0.1343	0.1348
220	0.1451	0.1446
240	0.1546	0.1534
260	0.1632	0.1617
280	0.1711	0.1696
300	0.1786	0.1771
320	0.1853	0.1845
340	0.1917	0.1916
360	(0.1983)	(0.1970)

temperature range between 3.08° and 4.05°K, ranges from 2.4 x 10^{-3} cal/g-$^{\circ}$K to 4.5 x 10^{-3} cal/g-$^{\circ}$K, with a maximum at 3.54° K. Morrison and Norton remarked that these values of specific heat are two orders of magnitude larger than those expected from elastic wave velocities of these samples. According to the Debye theory of solids, the specific heat at constant volume, C_v, at low temperature is given by

$$C_v = \frac{16}{15} \pi^5 \frac{k^4}{\rho h^3} \left(\frac{1}{v_p^3} + \frac{2}{v_s^3}\right) T^3 \tag{3}$$

where ρ is the density of the solid, k and h are Boltzmann's and Planck's constants, v_p and v_s are the compressional and the shear wave velocities, and T is the temperature in degrees Kelvin. For $v_p = 7$ km/sec and $v_s = 4$ km/sec of the glass spherules contained in both of the specimens 10017 and 10046, [16,17] the coefficient C_v/T^3 becomes 1.11 x 10^{-7} cal/g-$^{\circ}$K^4, whereas C_v/T^3 obtained from the experimental data are of the order of 10^{-5} cal/g-$^{\circ}$K^4. Morrison and Norton suggested the existence of an anomaly in the low-frequency range of lattice vibrational modes as a cause of the anomalous specific heat.

C. Thermal Conductivity

Since thermal conductivity k is related to thermal diffusivity $\varkappa$ by the relation

$$k = \varkappa \rho C_p \tag{4}$$

the data on thermal diffusivity can be used to estimate thermal conductivity if the specific heat at constant pressure C_p and the density ρ are known. In Fig. 3, thermal conductivity of type A and type C materials is calculated from the smoothed thermal diffusivities (1) and (2) by the use of relationship (4). Densities 2.95 g/cm^3 (type A) and 2.29 g/cm^3 (type C) are assumed to be constant over the temperature range considered. Since the specific heat of lunar samples seems to be similar regardless of the sample types, it is assumed that the specific heat of type C samples has the same temperature characteristics as type A samples.

For the same reasons mentioned in the section on thermal diffusivity, thermal conductivity of type C material is lower than that of type A. Even the type A material exhibits lower thermal conductivity than corresponding terrestrial basalt of nonporous texture. If the thermal conductivity of sample 10057 (type A) at 300°K, 3.56 x 10-3 cal/cm-sec-$^{\circ}$K, is corrected for the porosity of 0.174 by Maxwell's formula,

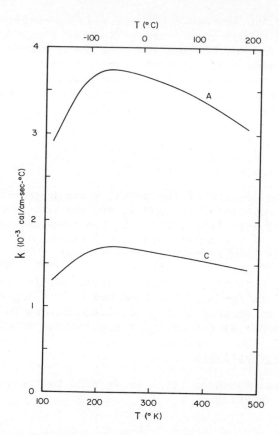

Fig. 3 Temperature variation of thermal conductivity calculated from thermal diffusivity and specific heat for
Apollo 11 type A and type C materials.

$$k' = [2(1-p)/(p+2)]k \qquad (5)$$

where k and k' are the corrected and the uncorrected thermal
conductivities and p the porosity, it becomes 4.68×10^{-3} cal/
cm-sec-°K. This value is not unreasonable for the conductivity
of nonporous basalt with high titanium content when compared
with the conductivity of ordinary terrestrial basalt.

It is expected that the lunar soil with much higher porosity
should exhibit lower thermal conductivities. The porosities
of the Apollo 11 drive-tube core samples are 46.5% (10005-0)
and 50.1% (10004-0), respectively.[1] Besides the effect of
porosity, the effect of contact resistance between the particles becomes more important in particulate materials like
lunar soil.

The data on thermal conductivity, directly measured on sample 10084 (lunar soil, type D), are available.[18,19] The measurements were made with a line heat source technique under the vacuum conditions up to 10^{-7} torr. The results are listed in Table 5. The bulk density of the sample was kept 1.265 g/cm^3 during the measurement. The conductivity of the lunar soil at a given temperature becomes independent of pressure below 10^{-2} torr.

Table 5   Thermal conductivity of Apollo 11 lunar sample 10084 (type D, lunar soil) under the pressure of 10^{-3} torr. Apparent density is 1.265 g/cm^3 [18,19]

Temperature, °K	Thermal conductivity, 10^{-6} cal/cm-sec-°K
	Type D (10084)
205	4.09
299	4.95
404	5.78

An intensive study of thermal conductivity was made on simulated lunar materials by Fountain and West.[20] It was shown that the thermal conductivity of particulate materials like lunar soils depends on bulk density (degree of compaction) and the pressure of gas medium filled in the specimen, as well as on other factors that control intrinsic thermal conductivity of dielectric solids. Fountain and West showed that the thermal conductivity of particulate materials is a function of temperature and is given by Watson's equation

$$k = \alpha + \beta T^3 \qquad (6)$$

where k is the thermal conductivity, T the temperature, and α, β are the constants. The second term on the right-hand side of Eq.(6) represents the contribution from radiative transfer which becomes dominant in particulate materials under high vacuum. The constants, α and β, depend on the bulk density of the specimen. In Table 6, the values of α and β, determined experimentally on the simulated lunar specimens, are shown as a function of density. The material used for this experiment is a basalt from the Columbia River, Ore., crushed to the size ranging from 37 to 62 μ. The measurement was made under the pressure of 10^{-6} torr.

The coefficients, α and β, determined by the least-squares from the data on sample 10084 (Table 5), are $\alpha = (3.99 \pm 0.26)$

Table 6   Coefficients α and β for particulate Oregon basalt in the equation $k = \alpha + \beta T^3$, where k is the thermal conductivity measured under the pressure of 10^{-8} torr and T the temperature[20] (units are changed)

Density, g/cm^3	Coefficients	
	α, 10^{-5} cal/cm-sec-$^\circ$K	β, 10^{-13} cal/cm-sec-$^\circ$K
0.79	0.122	0.404
0.88	0.155	0.399
0.98	0.142	0.411
1.13	0.212	0.454
1.30	0.296	0.581
1.50	0.392	0.820

$\times 10^{-6}$ cal/cm-sec-$^\circ$K and $\beta = (2.81 \pm 0.63) \times 10^{-14}$ cal/cm-sec-$^\circ$K^4.

A more detailed discussion on the experimental studies of simulated lunar materials will be given elsewhere in this volume.[25]

In the course of their measurements of specific heat, Morrison and Norton estimated thermal conductivity of samples 10017 (type A) and 10046 (type C) at liquid helium temperature.[15] The thermal conductivity was calculated from the time required for the specimens to attain thermal equilibrium. The result for sample 10046 was $2.5 \pm 0.5 \times 10^{-6}$ cal/cm-sec-$^\circ$K at 4°K. Sample 10017 was roughly estimated as 10 to 100 times more conductive than sample 10046.

Morrison and Norton remarked that, if the temperature dependence of thermal conductivity is negligible, thermal conductivity of type C material should be as low as type D material and is of the order of 10^{-6} cal/cm-sec-$^\circ$K at higher temperatures. This inference is substantially different from the estimation of thermal conductivity of type C material made by Horai et al.[6,7] according to which it is of the order of 10^{-3} cal/cm-sec-$^\circ$K. As indicated in Fig. 3, the thermal conductivities of type A and type C materials decrease toward zero as the temperature approaches to 0°K. It seems that the values reported by Morrison and Norton do not entirely contradict the trend suggested in Fig. 3.

Bastin et al.[13,14] also made thermal measurements on type A, type C, and type D materials. The values of thermal conductivity, provisionally obtained by them (supposedly at room temperature), are 0.5 to 2.0 x 10^{-3} cal/cm-sec-$^\circ$K for type A and type C materials, and 2.5 x 10^{-6} cal/cm-sec-$^\circ$K for type D material. A description of the method, together with the detailed discussion of the result, will be given by them elsewhere.

D. Thermal Parameter

Thermal parameter (the reciprocal of thermal inertia), which is defined by

$$\gamma = (k \rho C_p)^{-1/2} \tag{7}$$

where k is the thermal conductivity, ρ the density, and C_p the specific heat, is an important quantity that controls the variation of surface temperature of the moon. Thermal parameter γ can be calculated from the thermophysical quantities appearing in the right-hand side of Eq. (7). It is interesting to calculate the thermal parameters from the thermophysical data on lunar specimens because the thermal parameter of lunar surface material has been estimated from infrared and passive microwave observations from earth-based stations. In Fig. 4, thermal parameters of type A and type C materials are calculated from the thermal conductivities, densities, and specific heat. A composite effect of temperature on thermal conductivity and specific heat determines the temperature characteristics of thermal parameter. The thermal parameter decreases monotonically with temperature, and type C material exhibits higher thermal parameter than type A material over the entire range of temperature. The thermal parameter at 300°K, which is approximately the lunar equatorial mean surface temperature, is 23.3 cm^2-sec$^{\frac{1}{2}}$-$^\circ$K/cal (type A) and 38.9 cm^2-sec$^{\frac{1}{2}}$-$^\circ$K/cal (type C). These values are substantially lower than the estimate, 750 to 1250 cm^2-sec$^{\frac{1}{2}}$-$^\circ$K/cal, made by the infrared and passive microwave observations.[21,22]

In Table 7, the thermal parameter of specimen 10084 (type D) was calculated from the thermal conductivity, density, and the specific heat. The thermal parameter of lunar soil is more than an order of magnitude larger than that of lunar crystalline rock and breccia and agrees satisfactorily with the result obtained by the infrared and passive microwave observations. Considering that the maximum penetration depth of microwave is of the order of 100 cm and that the regolith of a thickness of several meters, consisting mostly of fine particles like lunar soil, is the commonest feature of the lunar surface, it is not

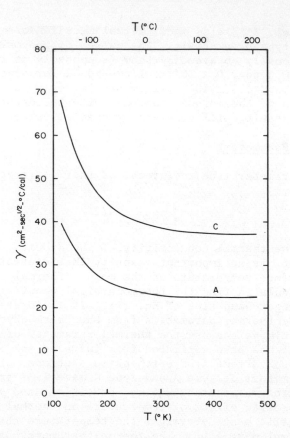

Fig. 4 Temperature variation of thermal parameter (the recip-
        rocal of thermal inertia) calculated from thermal con-
        ductivity and specific heat for Apollo 11 type A and
        type C materials.

surprising that the physical parameters of the lunar surface
material determined by the infrared and passive microwave
observations are mostly those of lunar regolith.

Table 7 Thermal parameter of Apollo 11 lunar sample 10084
         (type D, lunar soil)

Temperature, $^\circ$K	Thermal parameter, cm^2-$sec^{\frac{1}{2}}$-$^\circ$K/cal
	Type D (10084)
205	1198
299	950
404	811

III.   Rate of Heat Generation of Lunar Material
Returned by Apollo 11 and 12 Missions

One of the most important thermal properties of lunar material, the rate of heat generation, can be calculated from the
content of radioactive nuclides, ^{40}K, ^{232}Th, ^{235}U, and ^{238}U,
in the sample. The concentrations of potassium, thorium, and
uranium in eight Apollo 11 and eleven Apollo 12 samples were
determined by use of the low-background gamma-ray spectrometer
by the Lunar Sample Preliminary Examination Team.[1,5] The result of the determinations is summarized in Table 8. The rate
of heat generation of each nuclide is 0.765×10^{-8} cal/g-sec
for ^{40}K, 0.634×10^{-8} cal/g-sec for ^{232}Th, 0.136×10^{-6} cal/g-
sec for ^{235}U, and 0.225×10^{-7} cal/g-sec for ^{238}U. If the
terrestrial isotopic abundances are assumed, the rate of heat
generation of potassium, thorium, and uranium becomes $0.856 \times$
10^{-12} cal/g-sec, 0.231×10^{-7} cal/g-sec, and 0.634×10^{-8} cal/
g-sec, respectively. In Table 8, the rate of heat generation
is calculated for each sample from the concentrations and the
rates of heat generation of the radioactive elements. For
comparison, the concentration of radioactive elements and the
rate of heat generation for typical terrestrial rocks and
meteorites are shown in Table 9.[23]

The rate of heat generation of Apollo 11 crystalline rocks
(types A and B), with the exception of sample 10003, is not
significantly different from that of typical terrestrial basalt. Breccias and fines (types C and D) show slightly lower
rates of heat generation than crystalline rocks. The average
rate of heat generation for Apollo 11 crystalline rock, exclusive of sample 10003, is $0.386 \pm 0.027 \times 10^{-13}$ cal/g-sec. The
average for breccias and fines is $0.240 \pm 0.039 \times 10^{-13}$ cal/
g-sec.

Seven samples of Apollo 12 crystalline rock show remarkably
constant rates of heat generation with the average of $0.119 \pm$
0.017×10^{-13} cal/g-sec, which is considerably smaller than
that of Apollo 11 crystalline rocks. The Apollo 12 breccias,
fines, and feldspathic differentiate show higher rates of heat
generation which are comparable to those of terrestrial gra-
nitic rocks.

Figures 5 and 6 show potassium vs uranium and thorium vs
uranium relationships in Apollo 11 and 12 specimens. The K/U
ratio for Apollo 11 samples is 0.34×10^4, which is considerably lower than the ratio for chondrite, 8×10^4, or the average for terrestrial rocks, 1×10^4. An even smaller ratio,
0.12×10^4, is obtained for Apollo 12 samples. The Th/U
ratios are 4.1 for Apollo 11 samples and 4.0 for Apollo 12

Table 8   Contents of radioactive elements and rate of heat generation for Apollo 11 and 12 samples.

Specimen	Weight, g	K, %	Th, ppm	U, ppm	Rate of heat generation, 10^{-13} cal/g-sec
Apollo 11 samples					
Type A					
10057	897	0.242±0.036	3.4 ±0.7	0.78±0.16	0.417±0.058
10072	399	0.232±0.035	2.9 ±0.4	0.75±0.11	0.377±0.036
Type B					
10003	213	0.050±0.008	0.95±0.14	0.20±0.03	0.111±0.011
10017	971	0.227±0.034	2.9 ±0.4	0.70±0.10	0.365±0.034
Type C					
10018	213	0.144±0.022	2.3 ±0.3	0.60±0.09	0.297±0.028
10019	245	0.12 ±0.02	1.9 ±0.3	0.43±0.06	0.230±0.024
10021	216	0.120±0.018	1.8 ±0.3	0.39±0.06	0.215±0.024
Type D					
10002	302	0.11 ±0.02	1.6 ±0.3	0.46±0.10	0.217±0.030
Apollo 12 samples					
Crystalline rocks					
12002	1530	0.044±0.004	0.96±0.1	0.24±0.033	0.120±0.010
12004	502	0.048±0.004	0.88±0.09	0.25±0.033	0.118±0.010
12039	255	0.060±0.005	1.20±0.12	0.31±0.040	0.153±0.012
12053	879	0.051±0.004	0.89±0.09	0.25±0.033	0.119±0.010
12054	687	0.052±0.004	0.77±0.08	0.21±0.030	0.102±0.009
12062	730	0.052±0.004	0.81±0.08	0.21±0.030	0.104±0.009
12064	1205	0.053±0.004	0.88±0.09	0.24±0.035	0.116±0.010
Breccia					
12034	154	0.44 ±0.035	13.2 ±1.3	3.4 ±0.4	1.66 ±0.12
12073	405	0.278±0.022	8.2 ±0.8	2.0 ±0.3	1.0 ±0.09
Fines					
12070	354	0.206±0.016	6.0 ±0.6	1.5 ±0.2	0.745±0.060
Feldspathic differentiate					
12013	80	2.02 ±0.016	34.3 ±3.4	10.7 ±1.6	4.82 ±0.43

Table 9   Contents of radioactive elements and rate of heat
generation for typical terrestrial rocks and
meteorites[23]

Rock type	K	U	Th	Rate of heat generation, 10^{-13} cal/g-sec
Granite	3.79	4.75	18.5	2.58
Diorite	1.80	2.0	7.4	1.07
Basalt	0.84	0.6	2.7	0.38
Peridotite	0.0012	0.016	0.06	0.0076
Dunite	0.0010	0.001	0.004	0.0003
Eclogite (1)	0.036	0.048	0.18	0.023
Eclogite (2)	0.26	0.25	0.45	0.106
Chondrite	0.0845	0.012	0.04	0.012

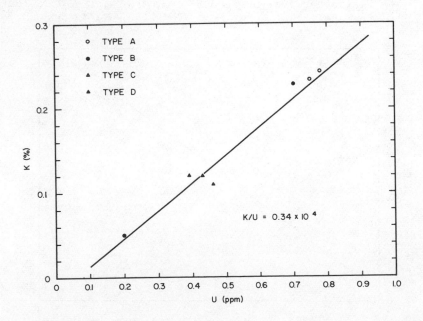

Fig. 5a   Potassium vs uranium in Apollo 11 samples. The correlation coefficient is 0.97. K/U = (0.34 ± 0.03)
x 10^4 (data from Ref. 1).

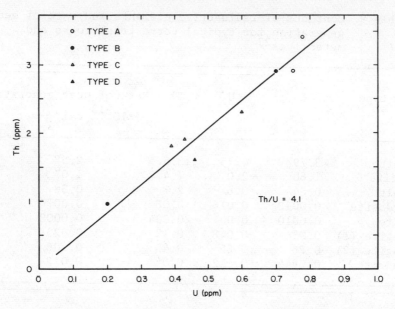

Fig. 5b   Thorium vs uranium in Apollo 11 samples. The correlation coefficient is 0.97.   Th/U = 4.10 ± 0.38 (data
from Ref. 1).

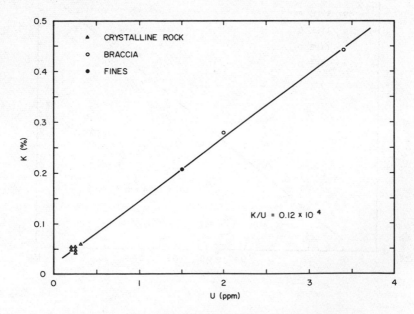

Fig. 6a   Potassium vs uranium in Apollo 12 samples. The correlation coefficient is 0.999.   K/U = (0.12 ± 0.002)
x 10^4 (data from Ref. 5).

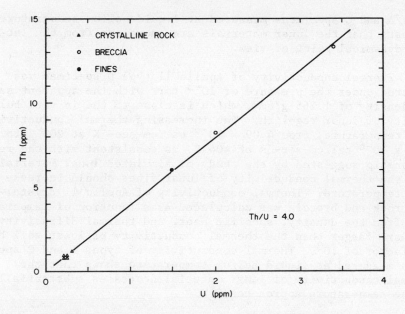

Fig. 6b   Thorium vs uranium in Apollo 12 samples. The correlation coefficient is 0.999. Th/U = 3.96 ± 0.04
(data from Ref. 5).

samples. They are not significantly different from chondritic
and terrestrial ratios.

IV. Summary

The data on thermal properties measured on the lunar materials returned by Apollo 11 and 12 missions were briefly reviewed:

1) Thermal diffusivity was measured on Apollo 11 type A
and type C samples in the temperature range between 150° and
440° K under atmospheric pressure. Thermal diffusivity of
type C material is lower and less temperature-dependent than
type A material. Both types of samples exhibit lower thermal
diffusivities than nonporous terrestrial basalt.

2) Specific heat of Apollo 11 type A and type D samples
was measured in the temperature range between 90° and 350°K.
The specific heat of the lunar specimens increases monotonically from 0.06 cal/g-°K at 90°K to 0.20 cal/g-°K at 360°K.
It is noteworthy that both types A and D specimens show almost
identical variation of specific heat with temperature regardless of the different mineral composition. Specific heat of

types A and C specimens measured at liquid helium temperature
suggests that the lunar materials are anomalous from the lattice dynamical point of view.

3) Thermal conductivity of Apollo 11 type D specimen was
measured under the pressure of 10^{-3} torr with the apparent sample density of 1.265 g/cm^3, which is close to the in situ bulk
density of lunar regolith. The increasing thermal conductivity
with temperature, from 4.09 x 10^{-6} cal/cm-sec-$^\circ$K at 205°K to
5.78 x 10^{-6} cal/cm-sec-$^\circ$K at 404°K, is consistent with the relationship suggested by the study of simulated lunar materials
that the thermal conductivity of lunar fines should increase
with temperature. Thermal conductivity of Apollo 11 crystalline rock and breccia was calculated as a function of temperature from the density, specific heat, and thermal diffusivity.
They are larger than the thermal conductivity of lunar soil by
the factor of 10^3. Thermal conductivity of types A and C specimens measured at liquid helium temperature shows that the
thermal conductivity of lunar material decreases substantially
as the temperature approaches 0°K.

4) Thermal parameter, the reciprocal of thermal inertia,
was calculated for Apollo 11 types A, C, and D materials from
the density, specific heat, thermal conductivity, and diffusivity. At 300°K, the respective thermal parameter of types
A and C material is 23 and 39 cm^2-sec$^{1/2}$-$^\circ$K/cal, whereas it is
950 cm^2-sec$^{1/2}$-$^\circ$K/cal for type D material. According to the
earth-based infrared and passive microwave observation, the
thermal parameter of lunar surface material ranges from 750 to
1250 cm^2-sec$^{1/2}$-$^\circ$K/cal. This may imply that the thermal properties that have been observed by the earth-based optical and
radiowave observations are those of lunar regolith.

Rate of heat generation of Apollo 11 and 12 samples was calculated from the concentrations of radioactive elements: potassium, thorium, and uranium. Apollo 11 crystalline rocks
show the average rate of heat generation, 0.386 x 10^{-13} cal/g-
sec, which is not significantly different from terrestrial
basalt. Apollo 11 breccias and fines show slightly lower rates
of heat generation. The average rate of heat generation for
Apollo 12 crystalline rocks, 0.119 x 10^{-13} cal/g-sec is considerably lower than that of Apollo 11 crystalline rocks.
Apollo 12 breccias, fines, and feldspathic differentiate show
the rate of heat generation, 0.75 to 4.8 x 10^{-13} cal/g-sec,
which is comparable to terrestrial granite. The K/U ratios
are 0.34 x 10^4 for Apollo 11 and 0.12 x 10^4 for Apollo 12
specimens. They are considerably smaller than either the average for chondrites, 8 x 10^4, or the average for terrestrial
rocks, 1 x 10^4. The Th/U ratio, 4.1 for Apollo 11 and 4.0 for

Apollo 12, is not significantly different from the chondritic
and terrestrial averages.

Acknowledgement

We acknowledge Drs. J. W. Lucas and A. E. Wechsler for valu-
able suggestions. This paper was prepared when one of us (K.
H.) was staying at The Lunar Science Institute, Houston, Texas,
under the joint support of the Universities Space Research
Association and the National Aeronautics and Space Administration - Manned Spacecraft Center under Contract No. NSR-09-051-
001. G. S. thanks the financial support provided by the
National Aeronautics and Space Administration Contract No.
NAS9-8102. (This paper constitutes The Lunar Science Institute
contribution no. 22).

Appendix: Apollo 11 Lunar Rock Specimens Used for Measurements of Thermophysical Properties[4]

10017 (type A): Very fine-grained, vesicular to vuggy,
glass-bearing poililitic olivine basalt.
Size: 10 x 11 x 7.5 cm. Weight: 975 g. Color: medium gray.
Shape: rounded to subrounded with one flat surface, hemispherical.
Fractures: very few fractures parallel to surface; hard coherent rock.
Pits and glass: pits with glass linings are sparce, white
haloes remain and are 1-3 mm diam; no glass splash. The glass
in the pits can be flaked off easily. Speckled white on rounded surfaces.
Texture: very fine-grained, vesicular to vuggy, subophitic to
granular; vesicles are smooth-walled with no filling or late-
stage reaction products; cavities or vugs are 1-3 mm, usually
elongate but not orientated; one large phenocryst of olivine,
rectangular shape, not resorbed.
Mineralogy: pyroxene: 45-50%, dark brown, ~ 0.1 mm; plagioclase: 30-35%, clear to chalky white, ~ 0.4 mm; ilmenite:
15-20%, submetallic, ~ 0.1 mm; olivine: $< 1\%$, medium yellow
green; 1 phenocryst 5 x 2 mm.

10020 (type A): Fine-grained, vesicular to vuggy, ophitic
olivine basalt.
Size: 9 x 8 x 4 cm. Weight: 425 g. Color: medium gray.
Fractures: very few fractures; hard coherent rock.
Pits and glass: glass-lined pits are numerous on rounded side;
white haloes without glass center are also common; pits are
very shallow with the center glass area 1 mm and the halo about
3 mm in diameter; rounded surface speckled white.
Texture: fine-grained, vesicular to vuggy, holocrystalline,

ophitic; vesicles are spherical and smooth-walled, up to 1 mm
diam; vugs are irregular up to 3 mm with euhedral plagioclase
and pyroxene up to 0.5 mm.
Mineralogy: plagioclase: 40%, clear to white, tabular and
stubby crystals, ~ 0.2 mm; pyroxene: 40%, honey brown trans-
parent, stubby crystals, ~ 0.2 mm; ilmenite: 15%, submetallic,
~0.3 mm; olivine: 5-10%, pale yellow green (chalky green
where subjected to impacts); phenocrysts, 2 to 3 mm.

10046 (type C): Breccia.
Size: 10 x 7.5 x 8 cm. Weight: 663 g. Density: bulk 2.45
g/cm^3. Color: dark gray.
Shape: equidimensional and rounded with two flat sides.
Fractures: numerous fractures, many throughgoing, rock is not
well indurated and friable.
Pits and glass: highly pitted on two rounded surfaces, other
four sides have a low density of pits. Largest pit 3 mm, pits
especially rich in beaded glass, but glass splashes are un-
common.
Texture: fragmental rock with fine-grained matrix enclosing
angular rock and mineral fragments; one highly vesicular fragment, 2 cm; one large glass sphere, 5-6 mm diam.
Mineralogy: feldspar and other mineral fragments identified.

10057 (type A): Fine-grained, vesicular to vuggy, granular
basalt.
Size: 10 x 11 x 6 cm. Weight: 919 g. Color: light gray.
Shape: triangular to trapezoidal shape, subangular, one subrounded surface.
Fractures: two sets of fine fractures, one coarse fracture
subparallel to rounded surface.
Pits and glass: evidence of pitting, some haloes and speckled
white feldspar; sparse yellowish-brown glass on surface; one
pit, 15 mm diam depression, 5 mm speckled white center.
Texture: fine-grained (0.05 - 0.12 mm), vesicular to vuggy,
granular; grain size coarser near vugs, vugs are irregular up
to 2 mm diam, plagioclase and pyroxene protrude into them;
vesicles are smooth-walled, 0.5-1.5 mm diam; inclusion of
nonvesicular crystalline fragment 5 x 2.5 mm with rounded
corners.
Mineralogy: plagioclase: 35-40%, clear to chalky white, good
cleavage; pyroxene: 45-50%, honey brown, stubby crystals;
ilmenite: 15%, submetallic, dark to black.

10065 (type C): Breccia.
Size: 10 x 7 x 5.5 cm. Weight: 350 g. Density: bulk 2.45
g/cm^3. Color: medium to dark gray.
Shape: hemispherical with one flat side.

Fractures: very few fractures, rock is well indurated, only
10056 is harder.
Pits and glass: pits with beaded glass on rounded surface,
medium density, 3 mm and smaller; on flat side very few pits
but two large (~3mm) ones are present.
Texture: fragments are coarser on the average in this breccia
than in other breccias.

References

[1] Lunar Sample Preliminary Examination Team, "Preliminary Examination of Lunar Samples from Apollo 11," Science, Vol. 165,
no. 3899, 1969, pp. 1211-1227.

[2] Fryxell, R., Anderson, D., Carrier, D., Greenwood, W. and
Heiken, G., "Apollo 11 Drive-Tube Core Samples: An Initial
Physical Analysis of Lunar Surface Sediment," Science, Vol.
167, no. 3918, 1970, pp. 734-737.

[3] Fryxell, R., Anderson, D., Carrier, D., Greenwood, W., and
Heiken, G., "Apollo 11 Drive-Tube Core Samples: An Initial
Physical Analysis of Lunar Surface Sediment," Proceedings of
the Apollo 11 Lunar Science Conference, Geochimica et
Cosmochimica Acta, Suppl. 1, Vol. 3, 1970, pp. 2121-2126.

[4] Schmitt, H. H., Lofgren, G., Swann, G. A., and Simmons, G.,
"The Apollo 11 Samples: Introduction," Proceedings of the
Apollo 11 Lunar Science Conference, Geochimica et Cosmochimica
Acta, Suppl. 1, Vol. 1, 1970, pp. 1-54.

[5] Lunar Sample Preliminary Examination Team, "Preliminary Examination of Lunar Samples from Apollo 12," Science, Vol. 167,
no. 3923, 1970, pp. 1325-1339.

[6] Horai, K., Simmons, G., Kanamori, H., and Wones, D., "Thermal
Diffusivity and Conductivity of Lunar Material," Science,
Vol. 167, no. 3918, 1970, pp. 730-731.

[7] Horai, K., Simmons, G., Kanamori, H., and Wones, D., "Thermal
Diffusivity and Conductivity, and Thermal Inertia of Apollo 11
Lunar Material," Proceedings of the Apollo 11 Lunar Science
Conference, Geochimica et Cosmochimica Acta, Suppl. 1, Vol. 3,
1970, pp. 2243-2249.

[8]Kanamori, H., Nur, A., Chung, D., Wones, D., and Simmons, G., "Elastic Wave Velocities of Lunar Samples at High Pressures and Their Geophysical Implication," Science, Vol. 167, no. 3918, 1970, pp. 726-728.

[9]Kanamori, H., Nur, A., Chung, D. H., and Simmons, G., "Elastic Wave Velocities of Lunar Samples at High Pressures and Their Geophysical Implications," Proceedings of the Apollo 11 Lunar Science Conference, Geochimica et Cosmochimica Acta, Suppl. 1, Vol. 3, 1970, pp. 2289-2293.

[10]Robie, R. A., Hemingway, B. S., and Wilson, W. H., "Specific Heats of Lunar Surface Material from 90 to 350 Degrees Kelvin," Science, Vol. 167, no. 3918, 1970, pp. 749-750.

[11]Robie, R. A., Hemingway, B. S., and Wilson, W. H., "Specific Heats of Lunar Surface Materials from 90° to 350° K," Proceedings of the Apollo 11 Lunar Science Conference, Geochimica et Cosmochimica Acta, Suppl. 1, Vol. 3, 1970, pp. 2361-2367.

[12]Robie, R. A., and Hemingway, B. S., "A Specific heat Calorimeter for the Range 4° to 400° K, Results for Calorimetry Conference Benzoic Acid," in preparation, 1970.

[13]Bastin, J. A., Clegg, P. E., and Fielder, G., "Infrared and Thermal Properties of Lunar Rock," Science, Vol. 167, no. 3918, 1970, pp. 728-730.

[14]Bastin, J. A., Clegg, P. E., and Fielder, G., "Infrared and Thermal Properties of Lunar Rock," Proceedings of the Apollo 11 Lunar Science Conference, Geochimica et Cosmochimica Acta, Suppl. 1, Vol. 3, 1970, pp. 1987-1991.

[15]Morrison, J. A., and Norton, P. R., "The Heat Capacity and Thermal Conductivity of Apollo 11 Lunar Rocks 10017 and 10046 at Liquid Helium Temperatures," Journal of Geophysical Research, Vol. 75, no. 32, 1970, pp. 6553-6557.

[16]Schreiber, E., Anderson, O. L., Soga, N., Warren, N., and Scholz, C., "Sound Velocity and Compressibility for Lunar Rocks 17 and 46 and for Glass Spheres from the Lunar Soil," Science, Vol. 167, no. 3918, 1970, pp. 723-732.

[17]Anderson, O. L., Scholz, C., Soga, N., Warren, N., and Schreiber, E., "Elastic Properties of a Micro-Breccia, Igneous Rock and Lunar Fines from Apollo 11 Mission," Proceedings of the Apollo 11 Lunar Science Conference, Geochimica et Cosmochimica Acta, Suppl. 1, Vol. 3, 1970, pp. 1959-1973.

[18]Birkebak, R. C., Cremers, C. J., and Dawson, J. P., "Thermal Radiation Properties and Thermal Conductivity of Lunar Material," Science, Vol. 167, no. 3918, 1970, pp. 724-726.

[19]Cremers, C. J., Birkebak, R. C., and Dawson, J. P., "Thermal Conductivity of Fines from Apollo 11," Proceedings of the Apollo 11 Lunar Science Conference, Geochimica et Cosmochimica Acta, Suppl. 1, Vol. 3, 1970, pp. 2045-2050.

[20]Fountain, J. A., and West, E. A., "Thermal Conductivity of Particulate Basalt as a Function of Density in Simulated Lunar and Martian Environments," Journal of Geophysical Research, Vol. 75, no. 20, 1970, pp. 4063-4069.

[21]Troitsky, V. S., "Investigation of the Surfaces of the Moon and Planets by Means of Thermal Radiation," Proceedings of the Royal Society of London (A), Vol. 296, no. 1446, 1967, pp. 366-398.

[22]Shorthill, R. W., "The Infrared Moon" (published elsewhere in this volume).

[23]Kaula, W., An Introduction to Planetary Physics: The Terrestrial Planets, John Wiley & Sons, New York, 1968.

[24]Kanamori, H., Fujii, N., and Mizutani, H., "Thermal Diffusivity Measurement of Rock-Forming Minerals from $300°$ to $1100°$ K," Journal of Geophysical Research, Vol. 73, no. 2, 1968, pp. 595-605.

[25]Wechsler, A. E., Glaser, P. E., and Fountain, J., "Thermal Properties of Granulated Materials" (published elsewhere in this volume).

THERMAL CHARACTERISTICS OF LUNAR SURFACE ROUGHNESS

Donald F. Winter*
University of Washington, Seattle, Wash.
John A. Bastin†
University of London, London, England
David A. Allen‡
Hale Observatories, Pasadena, Calif.

Nomenclature

A = albedo, dimensionless
a = radius of hemispherical crater, cm
a,b,d = linear dimensions of trough model, cm
B = Planck function, $W/m^2 \mu$
c = specific heat, J/g K
E = observed radiant energy flux, W/m^2 sec
f = fraction of surface covered with hemispheres in Pettit and Nicholson model
I = radiant energy flux, W/m^2 sec (vector or scalor)
k = thermal conductivity, W/m K
L_1 = thermal length scale for conduction, cm
L_2 = thermal length scale for radiation, cm
ℓ = scale of surface roughness, m or cm
$\underset{\sim}{n}$ = unit vector normal to local surface
$\underset{\sim}{p}$ = fraction of surface covered with high conductivity material
T = thermodynamic temperature, K
T_b = observed brightness temperature, K
T_L = theoretical Lambert temperature, K
T_s = surface temperature of plane surface model, K
ζ = local altitude of observer, degrees
θ = local selenographic orientation of length of trough measured clockwise from east, degrees
λ = wavelength, microns
μ = length, micron, 10^{-6} m
ξ = observer zenith angle, degrees
ρ = density, g/cm^3
σ = Stefan-Boltzmann constant, 5.675×10^{-8} W/m^2 K

---

*Faculty member, Department of Oceanography.
†Faculty member, Department of Physics, Queen Mary College.
‡Carnegie Fellow, Hale Observatories, Carnegie Institution of Washington and California Institute of Technology.

ϕ = local zenith angle of the sun, degrees
χ = local altitude of the sun, degrees
ψ = selenographic latitude, degrees
ω = angular frequency, sec^{-1}

I. Introduction

For many years thermal and microwave emission from the moon
has been used to deduce the thermophysical properties of the
lunar surface. It also has been recognized for a long time,
however, that the emissive properties of the surface are influenced by surface roughness. For this reason, infrared data
have been used extensively to obtain information concerning
lunar surface roughness, especially on a centimeter and millimeter scale. Not only is information on surface slopes of
importance, but so too is the scale size of the irregularities.
For example, very small-scale roughness, of magnitude less
than 10^{-4} m scale size, will have no effect on the thermal
properties other than to modify the effective albedo of the
surface. At this scale of roughness the surface will be almost
spatially isothermal. As we shall see in Sec. 2, two parameters which normally lie in the range $10^{+1} - 10^{-3}$ m serve to
specify the way in which the thermal properties of the lunar
surface depend on the roughness scale size.

Of course, various types of experiments or observations provide information about roughness of different scale sizes.
The principal methods of observation and associated scale-size
ranges are listed below, together with references: a) Earth-
based photography, $\ell > 10^3$ m; b) photographs from orbiting
vehicles, $\ell > 10$ m;   c) photographs from soft-landing vehicles, $\ell > 10^{-5}$ m; d) radar reflection, 1 m $> \ell > 10^{-2}$ m
(Refs. 1,2); e) photometric and polarization data of scattered
sunlight, $\ell \sim 10^{-6}$ m (Refs. 3,4,5,6); and e) infrared and
microwave emission, 10 m $> \ell > 10^{-3}$ m. In this chapter we
shall restrict our considerations to the last category and
discuss the relationship between the centimeter- and meter-
scale roughness of the surface and its emissive properties at
micron-wavelengths.

The reason for the dependence of lunar thermal characteristics on surface roughness is indicated schematically in
Fig. 1. Referring to the figure we observe that, in general,
any surface element at P will receive radiation not only from
the sun, but also from other surface elements. Furthermore,
the angle between the insolation vector $\underset{\sim}{I}_1$ and the normal $\underset{\sim}{n}$ to
the element at P will determine the rate at which the sun
irradiates a unit area of the surface. Both of these effects,
direct insolation and surface reradiation, need to be taken

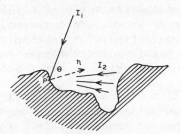

Fig. 1   Schematic diagram of intermediate scale lunar surface
         roughness.

into account in formulating the equation describing the heat
balance at P. This equation may be written in the form of a
surface boundary condition at P:

$$-k\ \underset{\sim}{n}\cdot\nabla T\ =\ (1-A_1)I_1 +\ (1-A_2)(I_2-\sigma T^4)\tag{1}$$

In this equation, I_1 is the solar energy directly incident per
unit time per unit area at P, and I_2 is the radiation from all
regions of the surface which can be viewed from P. A_1 and A_2
are the mean albedoes for radiation at solar and lunar temperatures. Finally, it has been assumed in writing down Eq.(1)
that reflected sunlight can be neglected. Within the surface
material itself it is usually assumed, as a first approximation,
that the transfer of heat energy is adequately represented by
the diffusion equation

$$\rho c(\partial T/\partial t)\ =\ \nabla\cdot(k\nabla T)\tag{2}$$

where ρ and c are the density and specific heat of the material.

When the thermal and electromagnetic parameters can be specified, Eq. (1) and (2), together with a description of the surface profile, can be used in principle to determine the time
variation of temperature at any point on or within the lunar
surface. In practice, however, the complexity of the necessary
computational techniques, as well as the inadequacy of our
detailed knowledge of the profile of the lunar terrain, severely limit any attempt to describe the situation exactly.
This chapter summarizes various attempts which have been
made to produce simplified models that can realistically describe the way lunar surface properties are modified by the
effect of the terrain. These attempts have proved successful
in that they predict results which are in very much better
agreement with observations than can be obtained from any
smooth surface model. Furthermore, the development of these
models made it possible to estimate the degree of roughness on

a centimeter and millimeter scale at a time when such information was not directly known. More recently, the study of these models in conjunction with near-infrared filter spectroscopy of the moon has revealed the existence of regions which are covered by a high density of meter-scale boulders.

Since the radiation diagram shown in Fig. 1 is a perfectly general one, it can be applied to roughness represented by any one of several types of lunar terrain. In particular, it can be used to refer either to the effects of (centimeter and meter scale) craters or to the effects produced by rocks lying on the lunar surface. In the case of craters, the picture needs simply to be specialized to represent depressions in the lunar surface material. For small craters of diameters less than 10 m, we may regard the whole crater as existing within the regolith layer and ascribe to this layer a thermal conductivity of the order of 10^{-3} Wm^{-1} K^{-1}. In the case of individual rocks lying on the lunar surface the picture is necessarily more complicated and the conductivity of the positive relief will be of the order of 1 Wm^{-1} K^{-1} with an underlying supporting bed of much lower conductivity.

The use of high-speed computing techniques has recently made it possible to investigate such models, and their properties will be described in subsequent sections of this chapter. Prior to these efforts, which have been undertaken almost entirely in the last five years, the only quantitative study of the lunar thermal properties was that reported by Pettit and Nicholson in 1930.[7] Both their measurements and analysis concerned the lunar day for which they justifiably assumed that the term $-k$ $\underset{\sim}{n} \cdot \nabla T$ in Eq. (1) could be neglected. Their roughness model consisted of a plane horizontal insulating surface partially covered with spheres set far enough apart that no sphere stood in the shadow of another. Radiative heat exchange between spheres and between a sphere and the flat surface was ignored. With these drastic assumptions Pettit and Nicholson obtained the following expression for the flux E across the disk under full-moon conditions:

$$E = E_o \left\{ \sin X_o + f \left[(2/3 \sin X_o) - 1 \right] \right\} \qquad (3)$$

where E_o is the energy flux that would be received from the surface if the spheres were absent, f is the fraction of the surface covered by spheres, and X_o is the elevation angle of the sun and observer for the extended flat space on which the spheres are placed. Equation (3) has clearly to be compared with the expression which is appropriate to a smooth surface:

$$E = E_0 \sin \chi_0 \qquad (4)$$

Pettit and Nicholson[7] compared their experimental measurements
with both of these expressions, and obtained much better agreement with Eq. (3); they were able to deduce a value of
f = 0.32. Perhaps the most serious shortcoming of their model
is that shadowing effects were ignored. These become important
when the zenith angles of either insolation or observation become large, and it is just these situations in which deviations
between observations and the prediction of smooth models are
greatest. In the remainder of this chapter we discuss three
recent and more realistic models of the lunar surface and present the latest infrared data with which they are to be compared.

II. Rectangular Trough Model

In this section we will consider the properties of a rough
model which has only horizontal and vertical surfaces and which
was investigated by Bastin and Gough.[8] The model consists of
a large number of trough-like indented elements, as illustrated
in Fig. 2(a), the direction of each element being oriented at

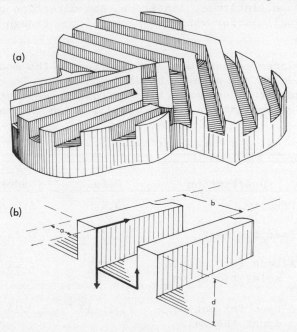

(a)

(b)

Fig. 2   Perspective diagrams of the rectangular trough model.

random with respect to the local lunar latitude and longitude.
All the top surfaces lie in one horizontal plane, and the
floor surfaces of the troughs lie in another. The linear
dimensions of all elements of the mosaic are large compared
with the trough dimensions a,b, and d shown in Fig. 2(b) and,
for this reason, shadowing effects at the boundaries between
the elements can be neglected. In any given calculation, the
trough dimensions were taken to be constant throughout each
element and for all elements of the array. The thermal and
radiative properties of models with different values of a,b,
and d were calculated in order to deduce the parameter combination which gave best agreement with observation. In the
calculations, insolation, emission from the surface, readsorp-
tion of emitted radiation, conduction, and the finite albedo
of the surface for incident and outgoing radiation were all
considered. The thermal and albedo parameters used in the
calculations are listed in Table 1. Although, for chronological reasons, these constants were determined entirely from
Earth-based observations, it is remarkable how little modifi-
cation they require as a result of the Surveyor and Apollo
programs. Adopting these physical constants, the model was used
to compute values of the brightness temperature in both the
mid-infrared and microwave regions. The calculations were
carried out for eclipse and lunation conditions as a function
of lunar phase, latitude, longitude, and direction of observation, as well as for various values of the trough dimensions
a,b, and d.

The calculated temperature distribution in the model natu-
rally depends on the size of the indentations. To characterize the size, it is convenient to introduce two thermal length
parameters or scales: $L_1 = (2k/\rho c \omega)^{1/2}$, which is the penetra-

Table 1   Physical constants adopted

Symbol	Description	Refs.	Value adopted
ρ	Density	3,4,9	1.0 g cm^{-3}
c	Specific heat	10,11,12	0.84 J g^{-1} K^{-1}
A_1	Albedo for incoming solar radiation	13,14	0.12
A_2	Albedo for outgoing lunar thermal radiation	15,16,17	0.14

tion depth of a thermal wave of frequency ω, and $L_2 = k/(\sigma T^3)$, which is the temperature scale height at a blackbody surface of temperature T during free radiative cooling. Throughout the lunation, for typical lunar surface material, L_1 is about 5 cm, while

$$L_2 \sim 4 \times 10^{-6} T^{-3} \text{ cm} \quad (T \text{ in K})$$

The role of subsurface conduction can be assessed by examining the relationships among L_1, L_2, and the roughness scale ℓ of the model. Thus, we can distinguish three regimes A,B, and C, defined as follows: A, $\ell \gg L_1$; B, $L_1 \gg \ell \gg L_2$; and C, $L_2 \gg \ell$. In regime A, the central interior regions of the raised portions of the model remain at an almost constant temperature. In regimes B and C, thermal effects can propagate across the ridges in a time short compared with ω^{-1} so that the temperature distribution within the ridges depends principally on the surface temperature. In cases where $\ell \gg L_2$, conduction at the surface does not significantly affect the energy balance. Thus the temperature distribution on the surface depends primarily on the surface geometry. Conduction is predominant in determining the surface temperatures in regime C when the ridges are virtually isothermal, and the surface is effectively smooth. During the night or an eclipse, conduction is important regardless of the scale of roughness. The results of a small sample of the calculations for lunation and eclipse conditions are shown in Figs. 3 and 4, for the case of a "standard" model with a = d = b/2 = 4 cm. Figure 3 shows isotherms within the soil at latitude 60° and at midday for various orientations θ of the trough. When $\theta = 0°$, the sun's radiation projected onto the cross-sectional plane is most glancing; the variations in surface temperature are most pronounced in this case. The maps for $\theta = 120°$ and 150° are almost identical to those for $\theta = 60°$ and 30°. Isotherms during and after an eclipse are shown in Fig. 4 for a region of the standard model with $\theta = 90°$ at the center of the lunar disk. The eclipse is one for which the trajectory of the sun's center in the lunar sky passes through the position of the Earth's center (maximum eclipse duration). Time is measured from the instant at which obscuration of the sun at the disk center begins. The surface temperatures T_s of the plane model are also shown. The last map corresponds to a time 30 minutes after the eclipse at the disk center has ended.

A large number of infrared observations of the moon are anomalous in the sense that they cannot be accounted for by a smooth homogeneous surface model whose thermal properties are temperature independent. By varying the parameters a,b, and d, good agreement can be established between the predictions of

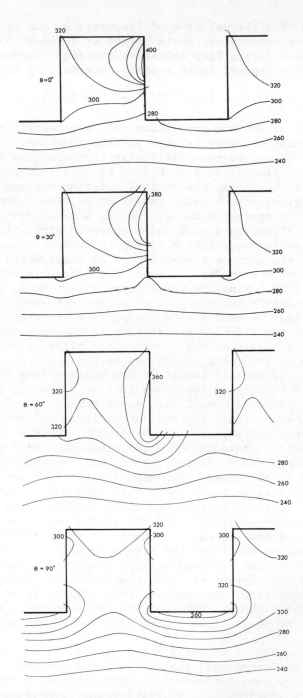

Fig. 3   Isotherms for regions of the standard trough model at
latitude ψ = 60°N at local noon.

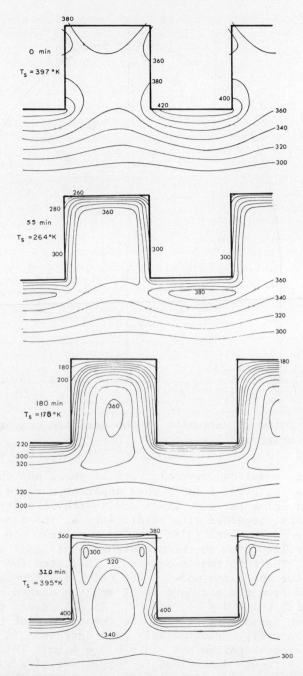

Fig. 4   Isotherms during and after eclipse for a region of the
standard trough model with θ = 90° at the center of
the lunar disk.

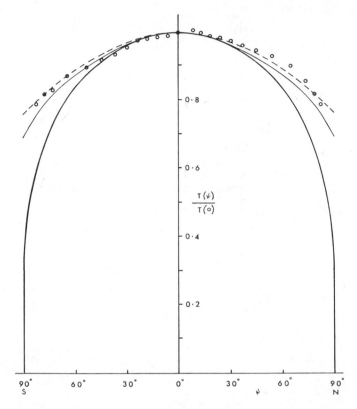

Fig. 5   Poleward variation of the meridian brightness temperature at 11 μ.

the rough model and several types of anomalous observations.
For example, Figs. 5 and 6 show apparent surface temperatures
found experimentally by Saari and Shorthill under full-moon
conditions, together with predictions at the same wavelength
from the rough model (also see Chapter 1.a).   In Fig. 5 the
open circles refer to the measurements of Saari and Shorthill.[18]   The thin continuous curve indicates the values obtained for the plane model.   The thick continuous curve was
obtained from the standard model and the dashed curve from the
model with b−a = d = b/4 = 4 cm.   In all cases, the results
have been normalized so that the temperature at the disk
center is unity.   Particularly good agreement is found for the
parameter combination b−a = d = b/4 = 4 cm.   This relationship
among the trough dimensions is suggestive of a surface which
is only moderately rough with one-quarter of the total area
consisting of crevices.   The same model is found to account
for the anistropy of surface radiation emitted from the sub-

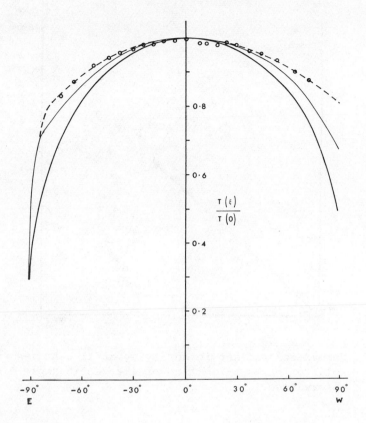

Fig. 6   Equatorial temperatures at full moon.

solar point, as shown in Fig. 7.   In this figure, open and
filled symbols refer, respectively, to values obtained from
the waxing and waning moon: triangles refer to Pettit and
Nicholson,[7] squares to Geoffrion et al.,[19] and circles to
Saari and Shorthill.[20]   The notation for the theoretical
curves is the same as in Fig. 6.

A further observation which can be explained by this roughness model is the preferentially rapid cooling of the lunar
limb during an eclipse.   Measurements reported by Sinton[12] and
by Saari and Shorthill[21] show that during eclipse, the drop in
brightness temperature at 11 μ near the limb compared with
that at the center of the disk is generally greater than would
be expected on the basis of a plane model.   The trough model
predicts an effect having the right order of magnitude; unfor-

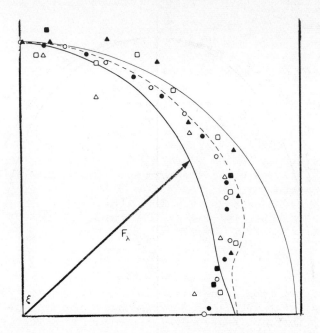

Fig. 7   Normalized radiant intensity F_λ at 11 μ at the subsolar point as a function of the zenith angle ξ of
observation.

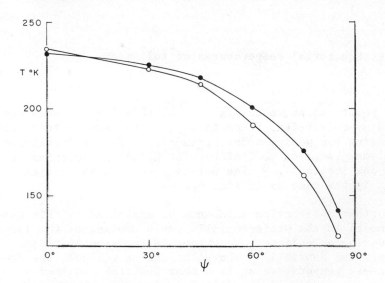

Fig. 8   Steady temperature below the penetration of the
thermal wave as a function of latitude.

tunately, the experimental data are not sufficient to permit
specification of a model with particular parameters.

As a final example of the application of the trough model,
we shall consider the observed poleward darkening of the
thermal microwave radiation from the moon. Figure 8 shows
as a function of latitude the predicted steady temperature
below the penetration of the thermal wave for rough and smooth
models. Open circles are the values calculated for the plane
homogenous model, closed circles are for the standard trough
model. The trough model predicts the greater poleward subsurface temperatures which are qualitatively in accord with
with microwave measurements. However, measured values generally exceed those predicted by the model.

III. Spherical Crater Model

The trough model is one of many possible geometrical formu-
lations of the moon's surface. In this section we will consider a somewhat different description of lunar roughness. We
know the moon to be peppered with craters of diameters ranging
from 10^6 to 10^{-4} m or less. Craterlets having diameters comparable with L_1 and L_2 could grossly modify the surface infrared emission. In order to put this hypothesis to the test,
we idealize crater interiors by spherical sections and their
environs by planes, as illustrated in Fig. 9. This model was
first proposed by Buhl, Welch, and Rea[22] to explain the full
moon limb brightening, reported by Pettit and Nicholson[7] and
Montgomery et al.[23] already illustrated in Fig. 5. Other
analyses based on similar models have been performed by
Winter and Krupp[24] and by Adorjan[25] who have investigated the
directional emission characteristics of sunlit equatorial
regions, over a wide range of sun and observer zenith angles.

An example of the data which inspired the spherical crater
model is exhibited in Fig. 10 which shows a collection of
apparent brightness temperatures of regions on or near the

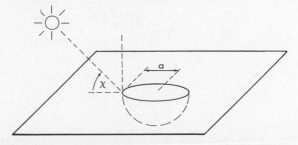

Fig. 9 Geometry of the spherical crater model.

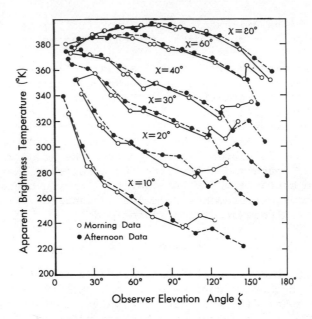

Fig. 10   Thermal meridian brightness temperatures as measured by Saari and Shorthill[18] at 11·μ. Observer elevation angle ζ and solar elevation angle X are illustrated in Fig. 11.

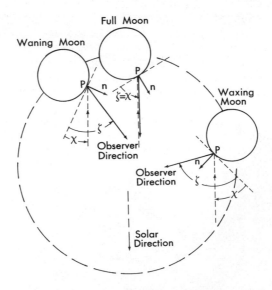

Fig. 11   Geometry of sun and observer elevation angles for a region on the thermal meridian.

thermal meridian, as measured by Saari and Shorthill.[18] The
thermal meridian is defined as the great circle through the
disk center and sub-solar point, and is approximately coinci-
dent with the lunar equator. We define in Fig. 11 a coordinate
system ζ, X, where ζ is the local altitude of the observer and
X is that of the sun. The observed temperatures are displayed
as functions of observer angle ζ for six different solar elevation angles. The angle ζ lies between 90° and 180° when
the sun and observer are on opposite sides of the local zenith
(see Fig. 11). An ambiguity is present in this convention:
the coordinates (ζ, X) are symmetrical about full moon, in
one case referring to a point lying east of the disk center,
in the other to a point to the west. In Fig. 10 the morning
and afternoon temperatures are distinguished by open and
filled symbols, respectively. The apparent temperature of a
smooth (Lambert) surface is independent of observer angle.
Figure 10 therefore illustrates the directional emission
characteristics of the lunar surface.

It is convenient to consider the ratio of the observed
brightness temperature T_b to the theoretical Lambert temperature T_L. The latter is a good approximation to the daytime
temperature of lunar material in the absence of surface roughness, but does not take into account the effects of the thermal
inertia of the surface at very low X. The ratio T_b/T_L (referred to subsequently as the D factor) is an indication of the
severity of local roughness.

All the important features of directional emission characteristics are displayed by the general pattern of D factor contours in the ζ-X plane. The contours shown in Fig. 12 were

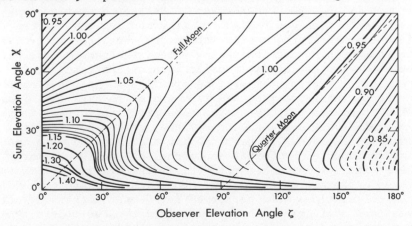

Fig. 12 Contours of averaged thermal meridian D factors
         (ratio T_b/T_L) in a ζ-X plane.

constructed from averaged, albedo-corrected equatorial data of
Saari and Shorthill. The data from which contours in the per-
ipheral zones $\zeta < 10°$ and $\zeta > 150°$ were drawn are somewhat
uncertain. Figures 10 and 12 show that the surface appears
anomalously warm when the sun is behind the observer ($\zeta = X$), the
emission being strongly peaked in the direction of illumination. On the other hand, the surface appears anomalously cool
to an observer who faces into the sun. From a qualitative
point of view, these features are consistent with the cratered
surface model. For example, those regions of crater interiors
which are illuminated more directly than the planar surroundings will generally be warmer than the environs. Shadowed
regions of crater interiors will be cooler than the environs.
It is reasonable to expect, therefore, that an illuminated
cratered surface will have just the directional emission we
seek.

As in the case of the trough model, radiation characteristics
of the surface are determined from surface temperature distributions. The geometry of the trough model allowed us to compute the temperature response for both a lunation and an
eclipse. This is not true for the spherical crater model,
however, and certain simplifications are necessary. To this
end, we shall confine our attention to thermal emission from
craters of scale ℓ, in the decimeter range or greater, under
daytime conditions. Throughout most of the day, when ℓ exceeds
both L_1 and L_2, conduction is relatively unimportant. In the
late afternoon and early morning, L_2 increases and conduction
begins to play a role in the determination of surface temperature. However, we are still justified in neglecting lateral
subsurface heat transfer around the crater when ℓ is greater
than both L_1 and L_2. Indeed, as a first approximation, subsurface conduction can be ignored in computing the daytime
surface temperature of crater interiors. Buhl et al.[22] have
shown that when conduction is neglected, the temperature distribution over the interior surface can be expressed in
closed form. Typical temperature profiles along the equatorial
arc of the crater interior are illustrated in Fig. 13. Buhl
and his coworkers calculated the equatorial brightness temperature profile at full moon for a plane surface dotted with
hemispherical craters. Their results are compared with measurements reported by Pettit and Nicholson[7] and Montgomery
et al.[23] in Fig. 14. The solid curves in the figure correspond
to three different areal surface densities of craters. Additional calculations were performed for deeper craters idealized
by vertical cylinders. The results from the two models together suggested that the observations could be matched by
postulating a surface with a 50% density of craters somewhat
deeper than hemispherical.

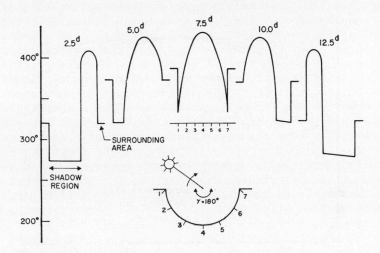

Fig. 13 Daytime temperature profiles in a hemispherical crater
         located on the thermal meridian. Time is measured in
         days after sunrise (Buhl et al.[22]).

When a correction is made to include conduction normal to
the surface, the temperature distribution must be found numer-
ically. Subsurface heat transfer in crater models has been
discussed by Winter[26] and Winter and Krupp.[24] In the latter
work, conduction effects have been incorporated in the analysis
in an approximate way to explain the equatorial data of Saari
and Shorthill,[20] which were exhibited in Figs. 10 and 12. Surface temperature distributions over crater interiors were calculated both for sharply contoured craters of hemispherical
configuration and for shallower craters, characterized by a
depth-to-diameter ratio of 1/4. These results were then combined with the thermal response of the planar environs to
obtain the total apparent brightness temperature of a cratered
soil illuminated at angle X, when viewed at angle ζ. The best
overall correspondence with measurement was achieved with comparable areal densities of shallow depressions, sharply contoured crater forms and planar environs. This result is in
accord with findings from in situ inspection of selected marial
regions by unmanned and Apollo spacecraft: high resolution
photographic studies and visual observations indicate that
local surface roughness is due, in large measure, to a broad
continuum of crater forms ranging from sharp to very subdued.

Comparisons of measurements with predicted temperature variations and D factor contours are illustrated in Figs. 15 and
16, respectively. The results exhibited in the figures

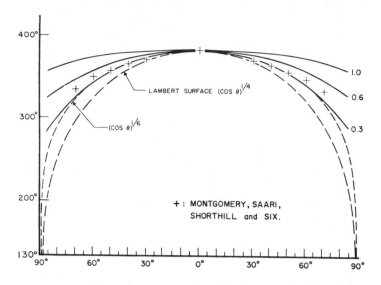

Fig. 14   Full-moon limb brightening (at 11 µ) for hemispherical craters on the thermal meridian for various
areal surface densities (see Ref. 22).

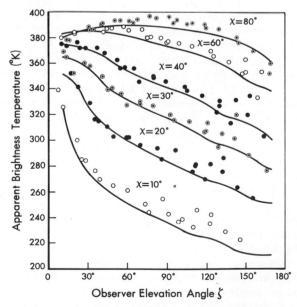

Fig. 15   Comparison of measured (11 µ) and calculated thermal
meridian brightness temperatures. The theoretical
temperatures were calculated using equal areal densities of plane surface, hemispherical craters, and
shallow craters with a depth-to-diameter ratio of 1/4.

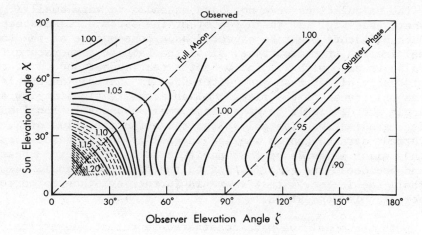

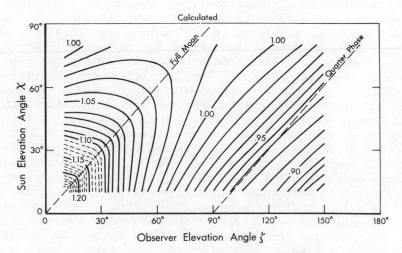

Fig. 16   (a) Contours of averaged thermal meridian D factors
          (ratio of T_b/T_L) in the ζ-χ plane from measurements
          by Saari and Shorthill.[18]
          (b) For comparison the D factor calculated by the
          spherical crater model. The crater configurations
          and areal densities are the same as in Fig. 14.

correspond to a nominal case with equal areal fractions of
planar environs, subdued and sharp craters. In certain ranges
of ζ and χ, somewhat better agreement with the data can be
achieved by postulating greater areal fractions of subdued
craters and fewer sharp ones. However, since the crater geometry is rather idealized, the model does not allow the iden-
tification of an optimum distribution of crater forms.

The calculations have not been extended to very small X, where (see Fig. 12) the D factor of the surface rises steeply. Such a feature might be expected as a consequence of the thermal inertia of the surface, since the Lambert temperature T_L falls to zero with X. The D factor near sunset should be considerably higher than just after dawn if thermal inertia is the only cause of the rise. Reliable observations at low sun angle are difficult to make since small changes in local surface gradient considerably alter the effective value of X. Current data suggest a smaller inequality than predicted and this might be evidence that positive relief features—rocks and boulders—are contributing to the observed signal. We shall see in Sec. 4 that there is further evidence to corroborate this suggestion.

Both the models we have considered represent geometrical idealizations of the real topography; each has its unique advantages and disadvantages. Although the spherical crater model describes a more realistic surface configuration, the trough model is more versatile and can represent either positive or negative relief. Despite the differences in geometry, however, the predictions of both models are qualitatively similar. Thus most of the important characteristics of infrared and microwave radiation from the sunlit surface of the moon can be explained by a surface which comprises a large proportion of level areas and of shallower, subdued features and a smaller fraction (1/3) of sharply depressed relief. But this is exactly the type of solution we find from the best fit trough model (b-a = d = b/4 = 4 cm) in which one-quarter of the relief is depressed. Thus we are given added confidence in our belief that the peculiarities of the moon's daytime thermal emission are caused by surface roughness. In Sec. 4 we shall examine the situation when the surface is not illuminated by the sun.

IV. Surface Relief and Hot Spots

It is well known that during an eclipse the surface of the moon displays hundreds of localised thermal enhancements. These same enhancements have been observed throughout the lunar night[27,28] (also see Chapter 3.b). Many different explanations of "hot spots" have been proposed in the literature over the past few years. These speculations are reviewed in detail by Winter.[26] Of the various hot spot mechanisms which have been invoked, the more plausible are: a) an increase in thermal inertia due to locally different thermal properties, an increase in soil particle size[29] or to an exposure of solid rock substrata[21,30]; and b) an increase in local relief associated

with surface roughness and/or superficial rocks and boulders.
Some or all of these mechanisms are undoubtedly operative in
varying degrees in different locations, but observations at a
single wavelength (or over an integrated wavelength range) cannot distinguish unequivocally between them. Differences in
the brightness temperature from region to region of the moon's
surface, measured at any one wavelength during the lunar night
or at eclipse, can be interpreted either as the result of
differences in the thermal conductivity of the soil or as the
result of a variation in the fraction of exposed or surface
rocks. It is true that we would not expect exactly the same
shape of cooling curve in these two cases, but the differences
are small and are less than present observational errors and
other uncertainties due to the variation of conductivity with
temperature and similar complexities. In this section we
shall attempt to elucidate briefly the role played by surface
relief in producing thermal enhancements during an eclipse and
by night.

During an eclipse, as well as a lunation, a surface which is
rough on a scale of centimeters or meters may appear anomalously warm due to "radiation trapping" in surface depressions.[8]
This effect was examined quantitatively by Bastin and Gough[8]
who studied the eclipse response of a number of roughness
models. They concluded that surface roughness is not the most
important cause of the more intense hot spots. Though it is
not difficult to produce a rough surface model which is as much
as 60 K higher in apparent temperature during eclipse than its
smooth surroundings, the same rough surface (with deep decli-
vities, for example) will also appear considerably hotter than
the smooth surroundings during the lunar day, and this is not
observed.

Further evidence comes from a study of conduction effects
during eclipse cooling of shallow, semicylindrical craters.[26]
During the umbral phase the central part of a warm crater interior tends to remain warm. At the same time, however, the
temperature of the immediate environs falls below that of
more distant surroundings owing to the combined effects of
lateral conduction and radiative loss from the vicinity of the
rim. The cooler annular regions tend to cancel the effect of
the warmer interior, thereby diminishing the apparent temperature differential of the hot spot. The conclusion seems
inescapable that while some of the less intense hot spots may
have a simple roughness explanation, the majority exhibit
thermal properties which can only be explained by a different
sort of roughness, such as positive relief.

The Apollo missions and their subsequent analysis of returned rock have shown that the surface regolith layer has a low conductivity ($\sim 10^{-3}$ W m^{-1} K^{-1}) but that lying on and imbedded within the surface are igneous rocks mainly in the centimeter and submeter size range whose conductivity is about a thousand times greater than the powdery fines which compose the regolith. It is still uncertain to what extent elsewhere on the moon there are appreciable areas of exposed rock of high conductivity or of surface rocks whose dimensions are in excess of a meter. Orbiter photography supports the suggestion that many surface regions have a high density of large surface rocks, as does an examination of Earth-based infrared observations—particularly those using a number of wavebands in the mid-infrared wavelength range. These measurements will be described and discussed below.

As pointed out earlier, variations in conductivity alone from one region to another on a flat lunar terrain will have little influence on the daytime surface temperature and will not normally give effects which are greater than those expected from local differences in the albedoes for incoming and outgoing radiation. However, calculations such as those performed by Krotikov and Shchuko[31] show that different values of conductivity predict quite different nighttime and eclipse temperatures; under both these latter conditions, regions of high thermal conductivity remain relatively hot. In the case of surface boulders or smaller exposures of high conductivity rock, we should again expect important effects on the eclipse and nighttime radiation, although in these cases the situation will be modified for the following reasons: 1) because of reradiation effects the net rate at which radiation leaves any surface of a rock will depend on the aspect of that surface and will be less than that from a plane horizontal surface at the same temperature; 2) for small rocks, less than about 20 cm in size, the rate of cooling will be more rapid than that of a smooth surface since the rock (in contrast to the plane homogeneous model) has a finite heat capacity which is quickly exhausted; and 3) surface exposures of rock of submeter size may lose appreciable amounts of heat by lateral conduction.

Roelof[32] made a detailed study of the rates at which surface rocks cool during eclipse and lunation conditions. The regime under discussion was essentially that of the roughness models discussed above, except that because of the high conductivity assumed for the surface rocks, the scale length parameters L_1 and L_2 were increased by factors of

$$\left[k_{rock}/k_{regolith}\right]^{1/2} \text{ and } \left[k_{rock}/k_{regolith}\right]$$

—i.e., about 30 and 1000 respectively. The condition that the surface temperature of a rock throughout the lunation should depend only on its directional aspect and its conductivity now requires that the rock be several meters in size. At eclipse, this limit is in the decimeter range.

The maximum rock size in Roelof's study was set, for various mathematical reasons, at 30 cm. The effect of a continuum of rock sizes was assessed by calculating overhead brightness temperatures of a rough surface characterized by two different rock size distributions. One was a fairly typical marial distribution inferred from Surveyor III photographs, whereas the other was the Surveyor I preliminary distribution with a flatter slope.[33,34] The latter surface was richer in larger debris and produced a significantly higher temperature enhancement. However, with the imposed particle size cutoff at 30 cm, the Surveyor I distribution could not account for the more intense hot spots.

This model was extended by Winter[26] to include larger rocks, temperature-dependent specific heat and an estimate of conductive heat exchange between the soil and the rock. The face temperature of the larger rocks was approximated by the surface temperature of an appropriately oriented semi-infinite solid. The effects of large rock populations were studied by using both the observed Surveyor VII rock distribution and a distribution with a slope intermediate between Surveyor VII and a marial "standard", as inferred from Surveyor spacecraft (see Fig. 17).[35] Both distributions produced typical hot spot brightness temperature variations. For example, Fig. 18 shows measured eclipse temperatures in the Kepler ray system anomaly, as reported by Shorthill,[36] together with theoretical results of Winter[26] using the intermediate rock distribution. In this calculation, the environs temperature includes a correction for directional effects that diminishes in importance during the eclipse. The umbral temperature enhancement above the environs is about 30 K. The Surveyor VII rock distribution produced an enhancement of 50 K, which corresponds to the measured differential of the most intense hot spots.

A systematic set of observations of the moon's mid-infrared radiation over a range of wavelengths has recently been reported by Allen[37,28] and Allen and Ney[38] who employed filters centered at wavelengths from 2.2 to 21 μ. They studied in particular the brighter eclipse hot spots. A differential technique was used in which the radiation from the enhanced region was modulated with respect to a point 52 seconds of arc away on the lunar disk. Their detector was a Ge:Ga crystal

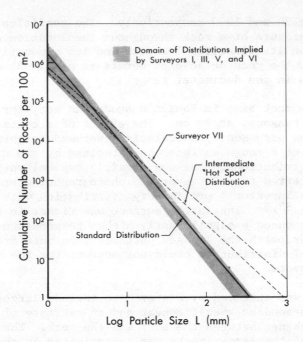

Fig. 17   Observed and hypothetical cumulative size distributions of exposed surface rocks.

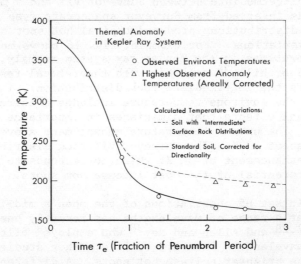

Fig. 18   Comparisons of observed (11 μ) and calculated brightness temperature variations of the Kepler ray system hot spot during eclipse. The measured temperatures were reported by Shorthill.[36]

with a square receiving area of 10^{-6} m^2 and was cooled to about
2 K. From the measurements of intensity a color temperature
T_c may be defined such that the intensity of lunar radiation
at two wavelengths, $I(\lambda_1)$ and $I(\lambda_2)$, is related to the Planck
function by the expression

$$I(\lambda_1)/I(\lambda_2) = B(\lambda_1,T_c)/B(\lambda_1,T_c)$$

Figure 19 shows data on the crater Tycho. The solid line
through each set of data is a blackbody curve at color temperature T_c. The figures accompanying each curve give, respectively, the cooling time since sunset in terrestrial days, T_c in
K and p, the proportion of the 26 second arc beam at the color
temperature. T_c is independent of wavelength to a high accuracy, although the radiation measured is much less than that
expected from a blackbody at temperature T_c. This is precise-
ly what we would expect from a composite body the majority of
which is at a low temperature but a fraction of which comprises small (i.e. unresolved) surfaces at a higher tempera-

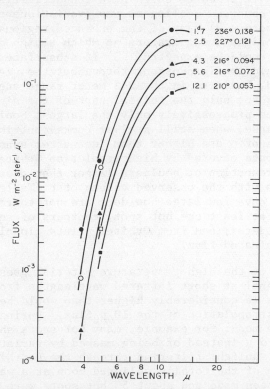

Fig. 19   Infrared photometry of the hot spot Tycho during the
          lunar night (see Ref. 39).

ture. By comparing the intensity measures with the blackbody function at temperature T_c, it is possible to deduce the fraction p of the surface covered with material of higher conductivity. p is found to vary from site to site, but is always much less than unity, Tycho being the most extreme example. At each hot spot measured, both p and T_c fall during the lunar night (Fig. 19). The use of differential chopping largely removes the effect of radiation from the cooler regolith, which in any case contributes only a small fraction of the total radiation.

This photometric technique has shown beyond doubt that the enhanced regions contain a small fraction of relatively high temperature surfaces during both an eclipse and the lunar night. There are reasons for preferring an explanation for the hot spots based on a model with boulders lying on the surface rather than with local exposures of high conductivity subsurface rock, and these we summarize below: 1) Rocks have been found on the lunar surface by both Apollo and Surveyor photography. Those returned to earth have conductivities consistent with the observed cooling curves (variations of T_c) in several hot spots. 2) Many of the enhanced regions are located around bright young craters at which a high density of surface rocks would be expected. 3) If subsurface rocks were exposed, p would remain constant throughout the night. On the other hand, boulders in the 2-5 meter range are at temperatures near T_c for only the initial part of the night, and p falls as we see progressively only the largest boulders. During an eclipse, when still smaller rocks contribute, the deduced values of p are higher than just after sunset. 4) For the few hot spots covered by high resolution Orbiter photographs, the proportion of boulders larger than about 2 m closely agrees with the observed values of p (Table 2). Regions seen to have few large boulders are not thermally enhanced. In at least one hot spot, contours of equal boulder proportion as determined from Orbiter prints closely match contours of infrared flux.

As a result of the high-temperature material present in hot spots, the signal at short infrared wavelengths from the enhanced regions is considerably higher than would be expected by a simple extrapolation of the 10 μ flux. During a total eclipse of the moon, for example, many hot spots should be detectable at 5 μ instead of being masked by variations in the scattered sunlight filtering through the Earth's atmosphere. A scan of the totally eclipsed moon at a wavelength of 4.8 μ has been made and about 50 hot spots were identified[40] having signals roughly in agreement with those predicted by the boulder hypothesis. Albedo effects seem to have been

Table 2 Boulder counts for crater hot spots

Crater	Fraction studied	Nominal resolution (meters)	Maximum boulder density, in 26" arc circle	Observed range of p, 26" arc beam	
Tycho	0.39	4.8	0.11	0.16 to	0.05
Aristarchus	1.00	2.8	0.057	0.06	0.02
Copernicus	0.38	2.3	0.050	0.06	0.015
Vitello	0.48	3.7	0.011	?	0.004
Dawes	0.31	2.5	0.003	0.003	0.001

comparable in this instance, and this may account for the sur-
prisingly weak signal from Mare Humorum, normally a prominent
hot spot.

It seems that there are sufficient boulders scattered across
the entire lunar surface to affect eclipse and nighttime measurements. Broad-band photometric data, for a region near the

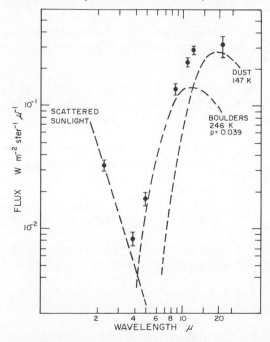

Fig. 20 Multicolor photometry of the south polar region of the
moon during the partial eclipse of Feb. 21st. 1970.

moon s south pole during a total eclipse, are plotted in Fig.
20. Three blackbody curves, for dust, boulders and scattered
sunlight, are needed to fit these data (see Ref. 37). This
region was selected to have no hot spots. Nevertheless, the
data imply contributions from both dust and rock, and at 10 μ
the boulders contributed two-thirds of the total signal.
"Temperatures" determined from a single waveband measurement
of the lunar surface by night or at eclipse are consequently
misleading.

We must conclude that while roughness can account for the
daytime thermal and microwave emission of the moon, in order
to describe the eclipse and nighttime observations we must
additionally consider positive relief features of high thermal
conductivity.

References

[1]Evans, J. V. and Hagfors, T., "On the Interpretation of Radar
Reflections from the Moon," Icarus, Vol. 3, 1964, p. 151.

[2]Pettengill, G. H. and Henry, J. C., "Enhancement of Radar
Reflectivity Associated with the Lunar Crater Tycho," Journal
of Geophysical Research, Vol. 67, 1962, p. 4881.

[3]Hapke, B. W., "A Theoretical Photometric Function for the
Lunar Surface," Journal of Geophysical Research, Vol. 68,
1963, pp. 4571-4586.

[4]Hapke, B. W. and Van Horn, H., "Photometric Studies of Complex Surfaces with Application to the Moon," Journal of Geophysical Research, Vol. 68, 1963, pp. 4545-4570.

[5]van Diggelen, J., "Photometric Properties of Lunar Crater
Floors," Recherches Astronomiques de l'Observatoire d'Utrecht,
Vol. 14, No. 2, 1959.

[6]Orlova, N.S., "Photometric Relief of the Lunar Surface,"
Astronomicheskii Zhurnal U.S.S.R., Vol. 33, p. 93.

[7]Pettit, E. and Nicholson, S.B., "Lunar Radiation and Temperatures," Astrophysical Journal, Vol. 71, 1930, pp. 102-135.

[8]Bastin, J. A. and Gough, D. O., "Intermediate Scale Lunar
Roughness," Icarus, Vol. 2, 1964, pp. 289-319.

[9]Salisbury, J. W. and Adler, J. E. M., "Density of the Lunar
Soil," Nature, Vol. 214, 1967, pp. 156-158.

[10] Garstang, R. H., "The Surface Temperature of the Moon," Journal of the British Astronomical Association, Vol. 68, 1958, pp. 155-165.

[11] Glaser, P. E. and Wechsler, A. E., "Small scale structure of the Lunar Surface," Icarus, Vol. 4, 1965, pp. 104-105.

[12] Sinton, W. M., "Temperatures on the Lunar Surface," Physics and Astronomy of the Moon, Chap. 11, edited by Z. Kopal, Academic Press, New York, 1962.

[13] Allen, C. W., Astrophysical Quantities, 2nd ed., Athlone Press, London, 1962, p. 145.

[14] Harris, D. L. Planets and Satellites, edited by G. P. Kuiper and B. M. Middlehurst, Univ. of Chicago Press, Chicago, Illinois, 1961, pp. 272-342.

[15] Burns, E. A. and Lyon, R. J. P., "Errors in the Measurement of the Temperature of the Moon," Nature, Vol. 196, pp. 463-464.

[16] Murcray, F. H., "The Spectral Dependence of Lunar Emissivity," Journal of Geophysical Research, Vol. 70, 1965, pp. 4959-4962.

[17] van Tassel, R. A. and Simon, I., "Thermal Emission Characteristics of Mineral Dusts." The Lunar Surface Layer, Materials and Characteristics, edited by J. W. Salisbury and P. E. Glaser, Academic Press, New York, pp. 445-468.

[18] Saari, J. M. and Shorthill, R. W., NASA Contractor Report, CR-855, 1967, Boeing Scientific Research Laboratories, Seattle, Wash.

[19] Geoffrion, A. R., Korner, M. and Sinton, W. M., "Isothermal Contours of the Moon," Lowell Observatory Bulletin, Vol. 5, 1960, pp. 1-15.

[20] Saari, J. M. and Shorthill, R. W., "Review of Infrared Observations," Physics of the Moon, Vol. 13, Science and Technology Series, American Astronomical Society, Tarzana, Calif., 1967, pp. 57-99.

[21] Saari, J. M. and Shorthill, R. W., "Isotherms of Craters on the Illuminated and Eclipsed Moon," Icarus, Vol. 2, 1963, pp. 115-136.

[22] Buhl, D., Welch, W. J. and Rea, D. G., "Reradiation and Thermal Emission from Illuminated Craters on the Lunar Surface," Journal of Geophysical Research, Vol. 73, 1968, p. 5281.

[23] Montgomery, G. C., Saari, J. M., Shorthill, R. W. and Six, N. F., Jr., Boeing Document D1-82-0568, 1966, The Boeing Company, Seattle, Wash.

[24] Winter, D. F. and Krupp, J. A., "Directional Characteristics of Infrared Emission From the Moon," The Moon, Vol. 3, 1971, p. 127.

[25] Adorjan, A. S., "Directional Behavior of Thermal Emission from a Rough Lunar Surface," AIAA Paper No. 71-480, AIAA 6th Thermophysics Conference, Tullahoma, Tennessee, 1971.

[26] Winter, D. F., "The Infrared Moon: Data, Interpretations, and Implications," Radio Science, Vol. 5, 1970, p. 229; and "Errata," p. 1497.

[27] Low, F. J. and Mendell, W. W., "Lunar Infrared Flux Measurements at 22 Microns," Bulletin of the American Astronomical Society, Vol. 1, p. 216.

[28] Allen,,D. A., "Infrared Studies of the Lunar Terrain, II," The Moon, Vol. 3 (in press), 1971.

[29] Winter, D. F., "Note on the Nonuniform Cooling Behavior of the Eclipsed Moon," Icarus, Vol. 5, 1966, p. 551.

[30] Fudali, R. F., "Implications of the Nonuniform Cooling Behavior of the Eclipsed Moon," Icarus, Vol. 5, 1966, p. 536.

[31] Krotikov, V. D. and Shchuko, O. B., "The Heat Balance of the Lunar Surface Layer During a Lunation," Soviet Astronomy AJ, Vol. 7, 1963, p. 228.

[32] Roelof, E. C., "Thermal Behavior of Rocks on the Lunar Surface," Icarus, Vol. 8, 1968, p. 138.

[33] "Lunar Surface Topography and Geology," in Surveyor I: A Preliminary Report, NASA SP-126, National Aeronautics and Space Administration, Washington, D.C., 1966.

[34] Shoemaker, E. M., Baston, R. M., Holt, H. E., Morris, E. C., Rennilson, J. J. and Whitaker, E. A., "Television Observations from Surveyor III," in Surveyor III, A Preliminary Report, NASA SP-146, National Aeronautics and Space Administration, Washington, D.C., 1967.

[35]Shoemaker, E. M., Baston, R. M., Holt, H. E., Morris, E. C., Rennilson, J. J. and Whitaker, E. A., "Television Observations from Surveyor VII," in Surveyor VII, A Preliminary Report, NASA SP-173, National Aeronautics and Space Administration, Washington, D.C., 1968.

[36]Shorthill, R. W., "The Infrared Moon: A Review," Journal of Spacecraft and Rockets, Vol. 7, No. 4, April 1970, pp. 385-398.

[37]Allen, D. A., "Infrared Studies of the Lunar Terrain, I," The Moon, Vol. 2, 1971, p. 320.

[38]Allen, D. A. and Ney, E. P., "Lunar Thermal Anomalies: Infrared Observations," Science, Vol. 164, 1969, pp. 419-421.

[39]Allen, D. A., "Infrared Studies of the Lunar Terrain." Ph.D. Thesis, University of Cambridge, 1970.

[40]Allen, D. A., Unpublished, 1971.

SECTION 4.   GEOPHYSICAL INTERPRETATION

Chapter 4a.   Thermal History of the Moon
Ray T. Reynolds, Peter E. Fricker, and Audrey L. Summers

CHAPTER 4a

THERMAL HISTORY OF THE MOON

Ray T. Reynolds,* Peter E. Fricker,[+] and Audrey L. Summers[†]

NASA Ames Research Center, Moffett Field, Calif.

Introduction

The interiors of planetary bodies have historically been
inaccessible to direct observation. Consequently, what knowledge we have of planetary interiors has primarily been dependent on theoretical calculations. Such calculations usually
rely on the construction of mathematical models based on known
physical principles and available experimental and observational data. Some of the Apollo 11 and 12 experiments, particularly the seismometer and magnetometer experiments, are capable of providing more direct information regarding the lunar
interior when the data are fully analyzed. Further experiments, including those of the lunar surface heat flow, should
also give more information on the state of the lunar interior.

Knowledge of the temperatures within the interior of a planet
is particularly difficult to obtain. Even for the earth,
where the density and pressure distributions are known rather
accurately, the temperature distribution is highly uncertain.
Since the effect of temperature on the equation of state for
materials within the earth is small compared with the effects
of pressure, the pressure-volume-temperature relations within
the earth must be much better known than at present to obtain
detailed temperature information from the experimental data.

We wish to thank Mr. Kenneth Claiborne for his assistance in
compiling data and assembling the bibliography. One of us
(P. E. Fricker) acknowledges partial support by the Aerospace
Institute Agreement between Ames Research Center and the
University of Santa Clara.
*Chief, Theoretical Studies Branch, Space Sciences Division.
[+]Currently Secretary General, Swiss National Science
Foundation, Bern, Switzerland.
[†]Research Scientist, Theoretical Studies Branch, Space
Sciences Division.

Many investigators have attempted to obtain the present
planetary temperature distributions by calculating thermal
evolutionary paths from mathematical models, beginning with
various initial conditions and employing detailed sets of
assumptions regarding the release, transport, and loss of
energy. Thermal history models for the moon have been calculated by Urey[60-62], MacDonald,[31-33] Levin,[26] Kopal,[24] Maeva,[36]
Phinney and Anderson,[41] and Fricker *et al.*[14]

Theories of Origin of the Moon

The problem of the thermal history of the moon cannot be
separated from the problem of the origin of the moon since the
thermal evolution must, to a large degree, be determined by
the conditions prevailing during formation. The major theories
of lunar origin fall generally into three groups: fission
theories, capture theories, and theories that envisage the
moon as forming in the vicinity of the earth as a double-
planet system. Each of these hypotheses has its own peculiar
strengths and weaknesses and it is not possible, at present,
to provide a definite proof for the correctness of any
particular one.

The fission hypothesis assumes that the moon was initially
part of a rapidly rotating earth. The earth then became rota-
tionally unstable and the moon was ejected. These theories
date from the early work of Darwin[11] and have been defended in
various modifications by others, including O'Keefe,[39] Wise,[66]
and Ringwood.[45] These theories provide an explanation of the
disparity in mean density between the earth and the moon by
supposing that the moon was derived from the earth's mantle.
Such theories have, however, been severely criticized on consideration of angular momentum. Most theories of this type
imply a molten or nearly molten earth at the time of fission.

Capture hypotheses postulate the formation of the moon in
another region of the solar system with subsequent capture by
the earth. Many calculations involving this type of capture
have been performed.[17,32,35,48,51] This theory explains the
earth-moon density difference in terms of formation of these
bodies in different parts of the solar nebula. The different
regions in which the earth and moon formed would be assumed to
have quite different iron to silicon abundance ratios. The
orbit of the moon must have then been changed sufficiently by
gravitational perturbations to permit its capture by the moon.
This hypothesis has been criticized on the grounds that the
probability of such a capture event is extremely low. There
are also problems of consistency with the present lunar orbit

(Goldreich[18]). Most calculations of this type seem to require a large tidal heat pulse near the time of capture by the earth.

The third hypothesis suggests that the moon formed by accretion from smaller bodies in the vicinity of the earth (Schmidt,[49] Ruskol,[48] Levin,[27] Opik,[40] and Ganapathy *et al.*[16]). Usually this process is considered to have occurred at the same time as the formation of the earth or shortly thereafter but some variations can provide for a delayed formative period of the moon. The major difficulty with this theory is the problem of explaining the large difference in the uncompressed mean densities of the two bodies. This discrepancy is due to a large difference in chemical composition between the earth and the moon. There is either a strong lunar depletion in iron relative to the earth or a relative enrichment of the moon in lighter elements. It has proven to be most difficult to provide an adequate physical mechanism for this fractionation process. This hypothesis does, however, avoid many of the major criticisms of the other two theories. Such a model will permit the formation of a body having low initial temperatures, although it does not rule out the possibility of a high-temperature origin.

There are many variations upon these general themes and many are being pursued vigorously at the present time (e.g., Urey and MacDonald[63] and Cameron[6]). For the purposes of calculating the subsequent thermal history of the moon, the theories of origin are of extreme importance since they determine the initial conditions for such calculations. Many of the fission theories, for example, imply extremely high temperatures during formation and thus would produce an initially molten and chemically fractionated moon. Some versions of the other two classes of theories can also involve high temperatures during the formative period. Early heating rates depend so strongly, however, on details of the tidal interaction process, accretion time, abundance of short-lived radioactives, and other, as yet poorly understood, processes that the possibility of formation of a homogeneous, low-temperature moon cannot be ruled out. The gross uncertainties involved in the formative process require consideration of the consequences of a wide variety of initial conditions.

Energy Balance

Before discussing the details of a thermal history calculation, it is useful to identify the major energy sources for the moon. Possible significant sources of energy within the moon, in addition to the specific heat due to the original temperature of the material within the protoplanetary nebula, include gravitational accretion, compression, decay of radioactive isotopes, electromagnetic interactions with the solar wind, and tidal interaction with other bodies.

The maximum energy resulting from gravitational accretion may be estimated for the moon by calculating the total gravitational potential energy for a lunar-sized body. Assuming a constant-density moon, the calculation results in an averaged energy total of about 1700 joules/g for the entire moon. The energy released per gram would, however, be considerably more for the outer layers and correspondingly less for the inner regions. The amount of energy that could be retained within a growing moon depends on the rate of accretion. Accretion rates for the moon are not known at present, and thus the amount of accretional energy retained is very uncertain.

The amount of compressional energy released in the interior of the moon by adiabatic compression during formation can be estimated easily and cannot amount to more than a few tens of degrees of temperature even at the center.

The decay of radioactive isotopes can produce a large quantity of energy during the lifetime of the moon. On the basis of their decay constants, the radioactive isotopes that are of possible importance as lunar heat sources can be divided into two groups. One group consists of the relatively long-lived radioactive isotopes U^{238}, U^{235}, Th^{232}, and K^{40}. The other group is composed of relatively short-lived extinct radionuclides such as Al^{26} (half-life, 72×10^6 yr), Be^{10} (half-life, 25×10^6 yr), I^{29} (half-life, 16×10^6 yr), and others. These isotopes could have produced large quantities of heat during the very early history of the moon. The energy produced by the long-lived radioactive isotopes over the lifetime of the moon has been estimated to be of the order of 1500-2000 joules/g, if the abundance of the radioactive elements within the moon are reasonably similar to those in the earth or in meteoritic material. The energy production from the short-lived radioactive isotopes is an indeterminate quantity since it depends on the initial abundance of these materials and, quite strongly, on the rate of accretion, both of which are unknown.

Sonett and Colburn[52] have proposed a mechanism for heating the interior of a planetary body with the electric current produced by a unipolar generator driven by the solar wind. The present energy flux of the solar wind is much too low to generate an appreciable heating rate by this mechanism. However, if the sun passed through a T Tauri stage of extensive mass

loss by a greatly increased solar-wind flow, their calculations indicate that the unipolar heating mechanism could have
extensively heated the lunar interior.

The heating of the lunar interior by tidal interactions
depends very strongly on the details of the evolution of the
lunar orbit and on the processes that dissipate tidal energy
within the moon.[23] Neither the dissipation processes nor the
course of orbital evolution are known to any satisfactory
degree. The amount and distribution of the energy contribution
by tidal interaction is thus very uncertain. Available estimates depend on the type of origin favored for the moon and
range from totally negligible amounts of energy to quantities
sufficient to melt the moon completely.

The energy released by various mechanisms will either
increase the internal energy of the moon or be radiated into
space. An increase in the internal energy will appear as an
increase in temperature or, if there are regions at the melting
point, as latent heat of fusion.

In spite of large uncertainties with regard to some of the
major energy sources, several simplifying assumptions can be
made that permit meaningful calculations. With the possible
exception of tidal energy release, the energy sources having
the largest uncertainties, that is, accretional energy,
electromagnetic heating, and decay of short-lived radionuclides
complete their contribution to the internal energy budget in a
time interval that is very short compared to the total lifetime of the moon. Thus the energy contributions from these
sources can be included implicitly, to a very good approximation, within the initial temperature distribution chosen for
the thermal history calculations. However, the energy produced by tidal friction cannot, in general, be considered in
this manner since it may not have been released during the
very early history of the moon. Until a definitive solution
of the problem of the orbital evolution of the earth-moon system is available, thermal evolution models are restricted to
two classes: one in which the tidal heating occurred early
enough to be included with the other sources of "initial"
energy, and another for which the contribution of tidal energy
to the energy balance is negligible.

To a first approximation, the thermal balance problem can be
expressed as one in which an "initial" or early temperature
distribution (the result of summing over a number of possible
energy sources) with an additional time-dependent energy contribution from long-lived radionuclides is modified by energy

transport processes and the loss of energy by radiation from the surface.

Planetary and Material Parameters

A number of parameters must be specified to calculate a lunar thermal history model. The numerical values used for radius, density, heat capacity, lattice conductivity, index of refraction, and surface temperature were adopted from Fricker *et al*.[14] (Table 1). Absorption spectra of silicate materials were investigated by Clark,[7] Kanamori *et al*.,[22] and others. On the basis of this work, a value of 20 cm^{-1} was used for the opacity. The assumed age of 4.6×10^9 yr for the moon is compatible with the age determinations for lunar material from the Apollo 11 and 12 landing sites.[1,2,56,58,59] The surface temperature was held constant at 0°C.

Although the abundances of the principal long-lived radioactive heat sources U, Th, and K have been determined in lunar samples from the Apollo 11 and 12 landing sites (see Ch. 3.b.III), uncertainties with regard to the initial concentration of these radionuclides in the lunar interior persist.

U contents of some 0.15–0.9 ppm, Th contents of about 0.4–4 ppm, and K contents of about 300–3500 ppm[10,19,21,28,50,56,64] have been measured in lunar Apollo 11 samples. The concentrations of U, Th, and K in crystalline basaltic samples from the Apollo 12 site are lower[29]; they fall into the lowermost part of the range of values given above for the Apollo 11 samples. On the other hand, the concentrations of radioactive elements in breccias, fine material, and a feldspathic differentiate from the Apollo 12 site are higher than the reported abundances in the crystalline Apollo 12 rocks and in the Apollo 11 rocks.[29]

The Th/U ratio of the investigated lunar samples is about 4 and thus is comparable to the Th/U ratio in terrestrial and

Table 1   Parameters for thermal calculations

Radius, r	1.738×10^8 cm
Density, ρ	3.34 g/cm^3
Heat capacity, c_p	1.2 joules/g-deg
Lattice conductivity, c	400 joules/g
Heat of fusion, H_f	7.89×10^5 joules/cm-yr-deg
Index of refraction, n	1.7
Opacity, ε	100 cm^{-1}
Surface temperature, T_s	273°K

meteoritic materials.[8,34] The K/U ratio, which is about 10^4
for terrestrial rocks and about 8×10^4 for meteoritic material,
is only about 2800 for the Apollo 11 rocks and about 2200 for
the crystalline rocks from the Apollo 12 site.

Even the relatively low abundances of radioactive heat
sources in the crystalline basaltic rocks from the Apollo 12
site are comparable to the concentrations in terrestrial
basalts and are much higher than the values that have been
estimated for the overall initial abundances of these radionuclides in the earth[20,30,34,42] and in the moon.[12,26] This
enrichment of radionuclides in the surface materials was
achieved by fractionation processes that must have been effective at least in the outer layers of the moon.

The earth's mantle and the moon are composed of silicate
materials having very nearly the same uncompressed mean density. For this investigation, the assumption was made that
the U and Th concentration in the moon was similar to that in
the earth's mantle. Wasserburg et $al.$[65] pointed out that the
K/U ratio is 1×10^4 for various groups of terrestrial igneous
rocks in comparison to a K/U ratio of about 8×10^4 for chondritic meteorites. Since the terrestrial rocks also have a
fairly uniform Th/U ratio of approximately 3.1, it was concluded that the average terrestrial material possesses a
higher U and Th concentration and a lower K abundance than the
chondrites. On this basis, Wasserburg et $al.$[65] estimated an
average U concentration of 3.29×10^8 g/g for the earth's mantle.
Other estimates for the average U concentration in the mantle
range from 2.8×10^8 g/g (model IV of Hanks and Anderson[20]) to
4.5×10^8 g/g.[34]

For the present calculations, the value of 3.29×10^8 g/g was
adopted for the average initial U concentration in the moon.
The measured abundances of radioactives in the Apollo 11 and
12 samples were taken into account by employing a K/U ratio of
2500 and a Th/U ratio of 4 (lunar abundances of Table 2).
Table 2 also contains the abundances of long-lived radionuclides for the Type III carbonaceous chondrites as proposed
by Urey and MacDonald[63] in their Table 11, assumption 4.

The melting temperatures in the lunar interior can be estimated from the melting behavior of terrestrial rocks and
meteorites provided the lunar material is of similar composition. Experimental investigations of melting temperatures of
silicate materials have been performed over the entire pressure range of 0 to 50 kbar that characterizes the lunar
interior.

Table 2 Abundances, heat generation, and decay constants
of radioactive heat sources

Radioactive nuclide	Abundance in 10^{-8} g/g				Radioactive heat generation, B_j, joules/g-yr	Decay constant λ_j $\times 10^{-10} yr^{-1}$[j]
	Chondritic[a]	Terrestrial[b]	Lunar[c]	Carbonaceous chondritic[d]		
U^{238}	1.092	3.067	3.27	1.84	2.97	1.54
U^{235}	0.0079	0.022	0.024	0.0133	18.0	9.71
Th^{232}	4.4	11.43	13.16	5.80	0.82	0.499
K^{40}	9.52	3.68	0.98	5.0	0.94	5.5

[a]MacDonald[31].
[b]Wasserburg et al.[65] and MacDonald[34].
[c]This paper.
[d]Urey and MacDonald[63] (Table 11, assumption 4).

The melting curves of terrestrial minerals including albite, diopside, and enstatite under such pressure conditions have been discussed by Fricker et al.,[14] McConnell et al.,[37] and others. The available data indicate that melting temperatures for multicomponent systems (e.g., basalt-eclogite) are generally lower than for individual minerals and should provide more realistic estimates of the melting conditions. The presence of such volatile constituents as H_2O and CO_2 within the melt will tend to lower the melting temperatures still further.

The Apollo 11 and 12 missions have shown that the lunar mare surface materials are partly composed of basaltic rocks. In view of their chemical composition, it seems unlikely that basaltic rocks constitute the major part of the moon. Evidence presented by Ringwood and Essene[46] and others support the hypothesis that the basalt formed by partial melting within the lunar interior. Experimental work by Ringwood and Essene[46] and by Akimoto et al.[3] indicates that the melting behavior of the Apollo 11 basalts is similar to that of anhydrous terrestrial basalts.[9,70]

Figure 1 shows the melting range of basalt from Cohen et al.[9] For the models discussed in this paper, the liquidus of basalt, which is very close to the solidus of peridotite (Kushiro et al.[25]), was adopted as the effective melting temperature.

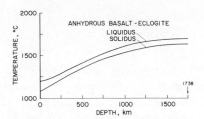

Fig. 1 Melting range of anhydrous basalt in the interior of the moon.

Thermal History Calculations

The internal temperature distribution for a solid moon with a constant spatial distribution of radioactive heat sources has been investigated for a wide variety of models by Urey,[62] MacDonald,[31] Levin,[26] Kopal,[24] Maeva,[36] and Phinney and Anderson.[41] Calculations by these authors generally give temperature distributions that increase with time and eventually exceed the melting temperatures for plausible lunar materials. The equation of heat conduction for a spherically symmetric solid body with variable internal heat sources is given as

$$\rho c_p \frac{\partial T}{\partial t} = \frac{1}{r^2} \frac{\partial}{\partial r} \left(r^2 k \frac{\partial T}{\partial t} \right) + A$$

where density ρ, heat capacity c_p, thermal conductivity k, rate of heat production per unit volume A, and temperature T are functions of radius r and time t. Boundary conditions that allow a solution to be determined are:

$$T(r,0) = f(r)$$

$$T(R,t) = g(t)$$

and

$$\left. \frac{\partial T}{\partial r} \right|_{r=0} = 0$$

The equation for thermal conductivity herein employed, from MacDonald[31] is

$$k = C + 16n^2 \sigma T^3 / 3\varepsilon$$

where C is lattice conductivity, n is index of refraction, σ is the Stefan-Boltzmann constant, and ε is opacity. The heat source term A may be expressed in the form

$$A = \sum_{j=1}^{\ell} B_j W_j \ e^{\lambda_j t} \rho$$

where the index j designates a specific radioactive isotope, ℓ is total number of such isotopes, B is rate of heat genera-tion per gram, W is mass fraction, and λ is the decay constant for each isotope.

A rectangular network in the rt plane is used to derive the finite difference equations approximating the partial differen-tial equation and the associated boundary conditions. The set of points in the rt plane given by $r = m\Delta r$, $t = n\Delta t$, where $m = 0,1,\ldots,M$ and $n = 0,1,2,\ldots$, is called a net whose mesh size is determined by Δr and Δt. It was determined that for an accurate and stable solution a selection of $\Delta r = 20$ km imposed a $\Delta t \leq 8 \times 10^6$ yr. The approximation to $T(m\Delta r, n\Delta t)$ is denoted by T_m^n. The finite differences chosen were

$$\frac{\partial T}{\partial t} = \frac{T_m^{n+1} - T_m^n}{\Delta t}$$

$$\frac{\partial T}{\partial r} = \frac{T_{m+1}^n - T_{m-1}^n}{2\Delta r}$$

$$\frac{\partial K}{\partial r} = \frac{K_{m+1}^n - K_{m-1}^n}{2\Delta r}$$

$$\frac{\partial^2 T}{\partial r^2} = \frac{T_{m+1}^n - 2T_m^n + T_{m-1}^n}{(\Delta r)^2}$$

which yield the difference equation

$$T_m^{n+1} = T_m^n + \frac{\Delta t}{\rho c_p} \left\{ \frac{1}{m(\Delta r)^2} \left(\left[\frac{m}{4} (K_{m+1}^n - K_{m-1}^n) + (m+1)K_m^n \right] \right. \right.$$

$$\left. \left. \cdot T_{m+1}^n - 2mK_m^n T_m^n - \left[\frac{m}{4} (K_{m+1}^n - K_{m-1}^n) - (m-1)K_m^n \right] T_{m-1}^n \right) + A_m^n \right\}$$

The boundary conditions become

$$T_o^n = T_1^n \ , \quad T_M^n = g(n\Delta t) \ , \quad n = 0,1,\ldots$$

and the initial condition is given as

$$T_m^o = f(m\Delta r) \ , \quad m = 0,1,\ldots,M$$

The results of representative calculations for this type of model (one that does not consider melting) are plotted in Fig. 2 for parameters taken from Fricker *et al.*[14] The parameters for this model are given in Tables 1 and 2. Figure 2 illustrates a "bulk" melting curve for lunar material as defined by Fricker *et al.*[14]

Effects of Melting on Thermal History Models

Because the onset of melting will greatly alter the subsequent lunar evolution, more realistic thermal models must consider the important effects of melting, including the energy required for latent heat of fusion, transport of heat by fluid convection in a molten region, and the movement of radioactive heat sources with time.

Heat of Fusion and Fluid Convection

Levin[26] discussed the importance of melting phenomena and calculated the conditions for the onset of melting in the moon. He accounted for the latent heat of fusion over the temperature interval between the liquidus and solidus by assuming a constant increase in the specific heat of the material over that interval. This assumption implies that the heat of fusion is absorbed linearly over the temperature interval. Reynolds *et al.*[42] demonstrated the importance of the consequences of melting and chose to define a stage of partial melting at the melting temperature, wherein the heat of fusion would be absorbed or released before the onset of complete melting or resolidification.

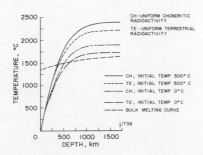

Fig. 2 Temperature distribution within a solid moon after 4.5×10^9 yr compared with a bulk melting temperature of silicate material.

The melting point gradient for possible lunar materials exceeds the adiabatic gradient in the lunar interior. Therefore, convection will occur within a molten region and, since heat flux in the interior is low and fluid convection is an extremely effective mechanism of heat transport, the internal temperatures will be prevented from exceeding the melting temperature by any appreciable amount. If the temperature significantly exceeds the melting point, convection will set in and the temperature will quickly be lowered to the vicinity of the melting temperature. An efficient convective process of heat transfer may be simulated by removing from the molten zone all thermal energy above that necessary to permit melting and depositing it in the adjacent outer radius interval.

Figure 3 shows the results of a thermal model calculation that includes the effect of absorbing the heat of fusion. Figure 4 shows a model that simulates fluid convection in the molten zone. Both of these models use parameters from Tables 1 and 2 and the bulk melting curve from Fig. 2.

Concentration of Radioactive Heat Sources

Another and even more significant consequence of the melting process involves the movement of radioactive heat sources. In the case of the earth, it is known that most of the radioactive isotopes, U^{238}, U^{235}, Th^{232}, and K^{40}, are concentrated in the crust. If the material from which the earth was formed was initially homogeneous, an extensive differentiation process must have occurred. It should be noted that the fractionation or differentiation mechanisms discussed here refer to those

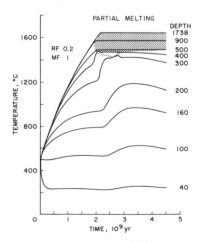

Fig. 3 Effects of melting on temperature distribution.

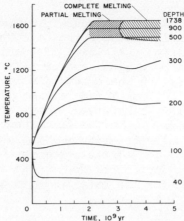

Fig. 4 Effects of melting and increased heat transfer in the
molten region on the temperature distribution in a
moon having terrestrial radioactivity and an initial
temperature of 500°C.

elements that provide significant radioactive heat sources.
In the event of melting, fractionation of other elements can
and does occur but it is not necessarily associated with any
large changes of density since the fractionation products are
silicates. The lunar iron abundance is too low to permit the
formation of a significant iron core.[43]

Although the record in the meteorites is not entirely understood, it is apparent that strong fractionation of radioactive
isotopes has also occurred within the meteorite parent
bodies.[15] The lunar samples returned from the Apollo 11 and 12
landing sites have confirmed the supposition that differentiation processes have also been operative within the moon. These
rocks are similar in many respects to terrestrial basalts and
contain comparable concentrations of U and Th. They are, however, lower in K.[28,29]

It is supposed that in the moon, as in the earth, the radioactive isotopes move outward with the lower melting fraction
during the differentiation process since the U, Th, and K
atoms with their large atomic radii do not fit well into the
silicate lattice structure. The outward movement of heat
sources is an extremely efficient method of heat transfer
because it removes not only the excess specific heat from a
given volume, but also all the heat energy that would have been
subsequently released by the decay of the radioactive material
that was carried out.

In the development of thermal model calculations, Fricker *et al.*[14] assumed that the movement of radioactives through solid high-density material is negligible, if operative at all, and that at least partial melting is required for an efficient fractionation process. In the event of melting, they allowed a given fraction of the long-lived radioactives, U, Th, and K, to move outward one radius interval during a single time interval. It was assumed that the fractionation process acted at the same rate for all the principal radioactive isotopes. The relation developed to calculate the mass fraction of a particular isotope as a function of radius and time, using the same notation as that for the heat conduction equation, is given by

$$W_m^{n+1} = \left\{ W_m^n + \Delta W_{m-1}^n \left[\frac{\rho_{m-1}^n (r_{m-1}^n)^2}{\rho_m^n (r_m^n)^2} \right] - \Delta W_m^n \right\} e^{-\lambda \Delta t}$$

where

$$\Delta W_m^n = \left\{ W_m^n + \Delta W_{m-1}^n \left[\frac{\rho_{m-1}^n (r_{m-1}^n)^2}{\rho_m^n (r_m^n)^2} \right] - RF \ e^{-\lambda t} \right\} MF$$

and Δt is the time step interval. The residual fraction (RF) of a given isotope identifies that mass fraction remaining within a given radius interval after melting allows the excess to move outward ($0 \leq RF \leq 1$). The moving factor indicates the fraction of the radioactives in excess of the residual fraction that is removed from a given radius interval in a single time step ($0 \leq MF \leq 1$). Detailed calculations of the effects of varying the MF and RF on temperature distributions were performed by Fricker *et al.*,[14] who concluded that an RF of 0.1 to 0.2 is consistent with geochemical data. A value of 1 was chosen for the moving factor, which expresses the assumption that the local differentiation process is efficient and operates on a smaller time scale than that of the movement of the melting front through the moon.

Figure 5 shows the results of a model calculation that include the effects of melting and the resulting concentration of radioactive isotopes toward the surface. The parameters are those from Tables 1 and 2, the bulk melting curve is from Fig. 2; RF = 0.2 and MF = 1. This and other differentiation models discussed by Fricker *et al.*[14] give a general trend of thermal evolution characterized by increasing temperatures in the outer layers early in the history of the moon. The period

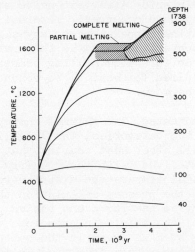

Fig. 5 Effects of melting and differentiation on the
       temperature distribution in a moon having terres-
       trial radioactivity and an initial temperature of
       500°C.

of rising temperatures is followed by a broad temperature maximum and, later in the lunar history, by a decline in the temperature of the outer parts of the moon. The time at which
this maximum is reached depends strongly on the initial temperature distribution. There are several general assumptions
made for the models and it is appropriate to discuss the significance of these assumptions in more detail.

The calculations assume no contribution to the thermal conductivity by convection until melting has occurred; after the
melting stage has been reached, the heat transport takes place
by means of fluid convection. This neglects the possibility
that large-scale convective motions may occur in the solid
state. Runcorn[47] has proposed this process for the moon and,
indeed, such a process is probably operative within the earth.
It is suggested here and by Fricker et al.[14] that the low temperatures in the outer layers and the small size of the moon
relative to the earth would minimize the importance of this
process. Studies of this process are in progress[5,57] and if
convective motions in the solid state are shown to transport a
significant amount of heat, thermal evolutionary models will
have to be modified accordingly.

Since melting, fluid convection, and differentiation are considered to occur at the melting temperature, the melting curve
for lunar material will determine approximately the thermal

state of the deep interior after extensive amounts of partial
or complete melting have occurred. Internal temperatures cannot exceed the melting temperature by any significant amount
because of fluid convection, and the temperatures in the deep
interior cannot drop far below melting temperature because of
the long times required for heat to be conducted from the
inner regions of the moon.

As previously indicated, the thermal energy sources have
been divided into two categories, the long-lived radionuclide
heat sources and all other heat sources that are collectively
assumed to be short-lived, with their energy contributions contained in the initial temperature distribution. This assumption excludes from the present calculation the possibility of
a large nonradioactive heating event taking place after the
formative period of the moon. In principle, there is no difficulty in including such an event in the calculations provided
the process can be accurately described in mathematical terms.

Influence of Initial Conditions on Thermal History Models

As the question of the lunar origin has not yet been
resolved, it is apparent that the thermal history of the moon
cannot be uniquely specified at present. There are, however,
a fairly large and, more importantly, a rapidly growing number
of clues that should eventually permit an understanding of both
these problems. Thermal model calculations can be used to
examine possible solutions and can aid substantially in the
analysis and interpretation of data from many sources, particularly with a view toward understanding the significance that
apparently unrelated facts may have toward an integrated and
fundamental understanding of the origin and evolution of the
moon. The range of possible models will be explored here by
first considering two cases that exemplify extreme sets of initial conditions and also illustrate the maximum differences
that may be expected between thermal history models. Each of
the two extreme models represents a case in which the movement
of radioactive materials with time does not affect the thermal
history. One model is assumed to be molten and completely differentiated at the time of formation. At the other extreme is
a model that was formed at temperatures so low that it did not
reach the melting temperature or concentrate its radioactives
at any time during its history. The parameters for these
models are given in Tables 1 and 2.

Initially Molten and Complete Fractionated Model

The initially molten model (model 1) is considered first. This model was assumed to be heated to the melting temperature throughout and to have all its radioactive heat sources concentrated in the outermost radius interval during the period of formation. Since the radioactive heat sources near the surface do not contribute significantly to the internal temperature distribution, this calculation involves only the heat conduction equation for a spherical body.

Such an initially molten and fractionated moon could be produced by any combination of initial heat sources including accretional heating, decay of short-lived extinct radionuclides, tidal interactions during the formative period, and electromagnetic heating. Differentiation of such an extensively heated body was considered to be effective enough to remove all but a negligible amount of the long-lived radioactive isotopes from the interior. The calculation was conducted for a time period of 4.6×10^9 yr and the melting curve employed was the liquidus of basalt-eclogite (Fig. 1). The variation of the temperature distribution with time for model 1 is given in Fig. 6, which demonstrates that, for this model, temperatures decrease monotonically with time.

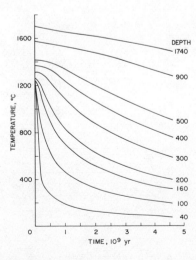

Fig. 6 Temperature distribution within an initially molten and completely fractionated moon (model 1); surface heat flow, 18.24×10^{-7} joules/cm^2-sec.

Homogeneous Model with Low Initial Temperature

To construct a nonmelting model (model 2) the initial temperature was assumed to be 0°C throughout the moon and the
radioactive abundances were adopted from Urey and MacDonald.[63]
Their abundances are given in Table 2 as representative of
values appropriate for a moon formed at low temperatures. For
a model that never reaches the melting temperature, the residual fraction is always unity and thus the moving factor is not
relevant. The results of the thermal model calculation were
plotted in Fig. 7, which also shows the melting conditions for
material of basaltic composition.[9] This calculation shows the
extreme difficulty in obtaining a lunar model that does not
achieve melting in the interior, even when the only heat
sources considered are long-lived radioactives. Temperatures
in this model come within a few hundred degrees of the melting
curve in the deep interior. The volatile materials, such as
water, that would be expected to be present within a moon
formed at low temperatures could depress the melting temperatures considerably below the anhydrous basalt melting curve
shown here.[25,69] It is possible that melting on a significant
scale could occur even within a moon having low initial
temperatures.

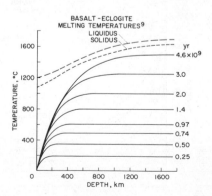

Fig. 7 Temperature distribution within an initially cold,
solid moon (model 2) having carbonaceous chondritic
radioactivity (Table 2); surface heat flow,
9.09×10-7 joules/cm2-sec.

Various mechanisms have been proposed that might reduce the abundance of potassium considerably by volatilization and thus reduce the amount of heat energy released from radioactive decay. These processes, however, imply a high-temperature origin and consequently might not reduce the peak temperatures achieved within the moon or decrease the possibility of the occurrence of melting.

Comparison of Extreme Models

The temperature distributions for the two extreme initial condition models tend to approach one another as the calculations proceed from the time of formation. In fact, the temperature distribution curves eventually cross and the temperatures within the low initial temperature homogeneous model actually exceed those within the molten and completely differentiated model. Because of the pronounced concentration of radioactive heat sources in the near-surface layers, the surface heat flow of the differentiated model is about twice that of the low initial temperature models 2 and 3.

Although new information may yet change our ideas regarding the thermal history of the moon considerably, our current knowledge is sufficient to permit some tentative conclusions. Both of the simple nondifferentiating models that have been discussed here have difficulties in meeting the present constraints on the lunar thermal history. This is particularly true for the low initial temperature model 2. If the evidence for extensive surface melting and differentiation at approximately the time of formation of the moon[67] is confirmed, the model with uniformly low initial temperatures must be ruled out. It is possible, however, that the heating source for this early surface melting event was limited to the outermost region of the moon. Most of the gravitational energy of accretion is released by the infall of material onto the outer layers. If the final stages of accretion occurred on a short-time scale, the accretional energy liberated could be sufficient to melt an outer layer of the moon.

Such an accretional surface melting process has been simulated by constructing a model similar to model 2 but having an initial temperature distribution such that the outer 200 km were molten and differentiated at the time of formation. The internal temperature distribution for this model (model 3,

Fig. 8) does not differ significantly from that of model 2
except for a rather short period following the initial accretional heating episode. Since an initially molten and differentiated moon would, of course, produce differentiated surface materials at or shortly after formation, both models 1
and 3 provide for an early surface melting episode.

The returned lunar samples have necessitated further constraints for thermal history models. Data from the rock crystallization age analyses indicate that melting occurred at the
Apollo 11 and 12 landing sites about 3.3 to 3.8 eons
ago.[1,2,58,59] These data have been interpreted to mean that
the maria floors were formed by massive lava flows from the
interior at the times given by the crystallization ages of the
rocks. It has been alternatively suggested that the maria
floors were formed of molten rock generated by large impact
events occurring a billion years or so after the formation of
the moon. However, in view of the morphology of the flow
structures and the crystallization history of the rocks,[4,44,46]
it appears more likely that the maria surfaces were formed by

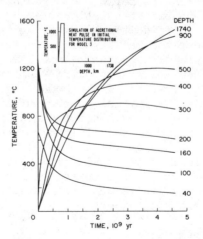

Fig. 8 Temperature distribution within an initially cold moon
       having an initially molten and fractionated outer
       layer 200 km thick and carbonaceous chondritic radio-
       activity (model 3); surface heat flow,
       9.65×10^{-7} joules/cm^2-sec.

lava flows of internal origin. A crucial test here involves
the spread between ages of different rock samples at a given
site. If there is a significant difference (order of 10^7-10^8
yr) between the ages of different specimens from the same
localized area, a single impact origin would be very difficult
to justify. The evidence is yet inconclusive.

A low-initial-temperature model, even with the addition of a
crustal melting heating episode as exemplified by model 3, cannot produce internally generated lava flows by the time of
maria formation indicated by the crystallization ages of the
lunar samples. This is because the outer layers of the moon
lose heat rapidly and are cooled well below the melting point
of silicates in a very short time.

In the case of the high-temperature, completely differentiated model it is also difficult to maintain sufficiently high
temperatures within the outer 200-300 km to provide internally
generated magmas at 10^9 yr after formation of the moon. The
outer layers of the high-temperature moon cool rapidly but,
contrary to the case of the low-initial-temperature moon, the
deeper interior can maintain relatively high temperatures.
Thus it is possible to hypothesize that a large maria-producing
impact event or large-scale tectonic movements could generate
fractures and zones of weakness that could reach to a sufficient depth to provide magma flows by pressure release. Such
a model would require the occurrence of the impacts or tectonic
movements at or near the time of production of the lava flows,
since the temperatures within the completely differentiated
model are monotonically decreasing with time, and the possibility of internally generated flows will consequently decrease,
also with time. These impact events would thus occur long
after the completion of the early accretional period of lunar
formation. The rapid cooling characteristic of this model
would predict that there are no internally generated surface
melting events significantly later than those that have already
been observed.

*Intermediate Initial Temperature Models with Time-Dependent
Fractionation*

The next level of complexity beyond the first two models considered involves models having intermediate values for the initial temperature distribution. Such models, having more than a
few hundred degrees of initial temperature, will reach the
melting temperatures in the interior during the 4.6×10^9 yr cal-
culational period with attendant redistribution of the radioactive heat sources. Fricker et al.[14] showed that a

characteristic feature of models of this type, which melt and permit the radioactive isotopes to move toward the surface, is an internal temperature distribution that rises to a maximum value in the outer layers and then begins to decrease.

The period of maximum heating within the outer layers would seem to be the most likely period for the generation of silicate magmas and their release to the surface. The initial temperature distribution can be chosen so that this period of maximum heating within the outer layers will coincide with the interval of maria formation. Wood[68] has recently performed calculations based on similar assumptions. This type of model, provided a crustal melting process similar to that discussed for model 3 is also included, can provide much more likely conditions for the internal origin of the maria lava flows. Such intermediate models are not unique, since the form of the initial temperature distribution is not known. It is possible, however, to calculate self-consistent models that embody the currently known constraints and can be considered to be representative of the present state of our knowledge.

Two simple forms for the initial temperatures were used to illustrate the behavior of this type of model. One is characterized by a constant initial temperature of 1250°C with near surface melting due to accretion. The other initial temperature distribution is assumed to be 300° below the melting curve for basalt, with the addition of a crustal heating event. The level of these temperature distributions was adjusted to provide maximum temperatures within the outer layers during the period of maria formation (approximately 3.3–3.8 b.y. ago). Results of calculations using these temperature distributions (models 4 and 5) are shown in Figs. 9 and 10. A moving factor of 1 and a residual factor of 0.1 were used for these models. The radioactive abundances are the same as for model 1 (Table 2). The radioactive heat sources are allowed to move when the temperature reaches the liquidus of basalt. The fraction of molten material at this temperature was assumed to be small and the heat of fusion negligible. The heat of fusion required upon melting of the bulk of the material would be significant but these temperatures are not reached after the heat sources are removed. The other parameters are the same as those of model 1. These model calculations give interior temperatures that are increasing rapidly in the deep interior and decreasing rapidly near the surface for the first 5×10^8 yr or so. Melting in the deep interior then begins, and radioactive heat sources begin to move outward. Temperatures in the deep interior then level off and begin to fall slowly as those regions are depleted in heat-producing radionuclides, while the outer

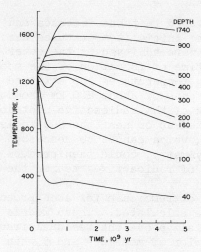

Fig. 9 Effects of melting and time-dependent fractionation of
radionuclides on the temperature distribution within an
initially hot moon having a constant initial temperature
of 1250°C and an initially molten outer layer 200 km
thick (model 4); surface heat flow, 17.79×10^{-7}
joules/cm^2-sec.

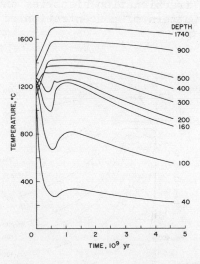

Fig. 10 Effects of melting and time-dependent fractionation of
radionuclides on the temperature distribution within an
initially hot moon having an initial temperature 300°C
below the basalt liquidus and a molten and fractionated
outer layer 200 km thick (model 5); surface heat flow,
17.84×10^{-7} joules/cm^2-sec.

layers become enriched in these isotopes and temperatures
begin to rise. The initial temperatures were chosen so that
peak temperatures were achieved in the outer layers at roughly
10^9 yr after the time of formation of the moon. For these
models the melting temperatures are attained in the lunar
interior to a distance of about 200 km from the surface, and
therefore the bulk of the radioactive heat sources finally
becomes concentrated at or near the 200-km level. Subsequent
igneous processes, which cannot be described by equations
assuming radial symmetry, could then produce local movement
and concentration of radioactives nearer the surface.

To simulate the extreme case for such processes, two additional models were calculated. These models assumed that all
radioactive materials above the residual fraction were carried
out from the interior of models 4 and 5 when the peak temperatures were reached at the level of maximum concentration of
radioactives. These models, 6 and 7 (Figs. 11 and 12), which
are based on models 4 and 5, respectively, provide for much
more rapid cooling of the outer layers. Thermal histories for
which the radioactive heat sources are concentrated at depths
between 200 km and the surface should be intermediate between
these two sets of models. The maria and highland areas appear
to have different fractionation histories and should certainly
have a different pattern of concentration of heat sources with
depth.

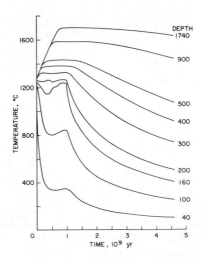

Fig. 11 Modified model 4 with extreme time-dependent fractiona-
        tion of radionuclides simulating maximum igneous activ-
        ity (model 6); parameters same as for model 4; surface
        heat flow, 18.50×10^{-7} joules/cm^2-sec.

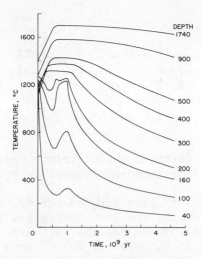

Fig. 12 Modified model 5 with extreme time-dependent fractionation of radionuclides simulating maximum igneous activity (model 7); parameters same as for model 5; surface
heat flow, 18.46×10^{-7} joules/cm^2-sec.

Discussion of Observational and Experimental Constraints
on Lunar Thermal History Models

In order for the interior to reach the melting temperature
and differentiate rapidly enough for the peak temperatures in
the outer layers to occur within a billion years or so after
formation, the initial temperatures within the interior must
be rather high, of the order of 1000°C or more. Partial or
total release of pressure in the interior by fracture would
permit the melting of material that is above its surface melting temperature. Figure 13 shows the depths at which the surface melting temperature is reached for both the time-dependent
fractionation models and the two extreme initial condition
models.

The low-initial-temperature models are unable to generate
internal magmas by pressure release during the period of 0.8
and 1.3×10^9 yr after origin. The completely molten and differentiated model would have to release magmas from considerably
greater depths than models of the type of 4 through 7. It
seems unlikely that rock ages from the two lunar sites currently explored define the limits of the period of lava generation. Earlier or later events, if discovered, would further
increase the difficulties for the two extreme models.

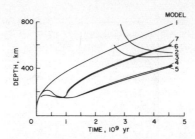

Fig. 13 Time vs depth at which surface melting temperature
(1200°C) of anhydrous basalt is reached.

Urey and MacDonald[63] have singled out two arguments as the
most compelling reasons for postulating low initial temperatures for the moon. These are the estimate of the average temperature of the moon inferred by Ness[38] from Explorer 35 magnetometer data, and the existence of nonhydrostatic conditions
within the moon as inferred from the figure and the presence
of mascons. Although the value reported by Ness has since
been discounted,[54] the investigation of the lunar interior by
the interpretation of surface magnetometer data holds great
promise for determining the present temperature distribution.
The establishment of a magnetometer on the moon's surface by
Apollo 12 made available, for the first time, measurements
that can probe the deep interior of the moon when taken
together with data provided by Explorer 35 in lunar orbit.
The electromagnetic excitation of the moon by fluctuations in
the solar wind produces electrical currents in the lunar interior. The magnetic fields associated with these currents are
detectable at the surface of the moon.[13] Analysis of the measurements of these fields can permit determination of the bulk
electrical conductivity profile of the lunar interior. Values
for the electrical conductivity within the moon have been
given by Sonett *et al.*[55] and Dyal and Parkin.[12] To obtain the
internal temperature distribution from the electrical conductivity profile, the conductivity-temperature relation for
lunar material must be known. At present this relation is not
well known and temperature estimates depend strongly on the
type of material assumed to comprise the moon. Sonett *et
al.*[53,55] and Dyal and Parkin[12] calculate internal temperature
distributions that vary from values much lower than those of
models 1-7 for high-conductivity materials such as iron
basalts to values approaching those given in models 1-7 for
low-conductivity materials such as olivine and peridotite.

A more precise knowledge of the present internal temperature
distribution would be of great value and could permit the
range of possible thermal history models to be narrowed

considerably. Unfortunately, a sufficient number of
undetermined parameters would remain and make it impossible to
reconstruct an unique solution for the thermal history even if
the present temperature distribution were known exactly.

The present surface heat flow of 18×10^{-7} joules/cm^2-sec for
an initially molten, differentiated model (model 1) is higher
by about a factor of 2 than the calculated heat flow for an
initially low-temperature, undifferentiated model (model 2)
even if an outer molten and fractionated layer of 200 km is
taken into consideration for the cold model (model 3). This
significant difference should permit a distinction of the two
possible extreme cases by means of direct heat-flow determinations on the lunar surface (see Chap. 2.c). The calculated
surface heat flow for models with time-dependent fractionation
of radionuclides (models 4-7) is similar to that of an initially molten and fractionated model. Thus, definitive clues
regarding the reconstruction of these two types of "hot"
models could probably not be obtained by lunar heat-flow measurements. On this basis, one could also predict that the surface heat flow from maria and highland areas will be rather
similar.

The argument that the moon's interior must be very cold in
order that the mascons and the figure remain out of hydro-
static equilibrium depends on the amount of structural strength
assumed for the lunar interior. The outer layers of the moon
are considerably colder and more rigid than comparable levels
within the earth. The load that can be supported within the
moon is not known, and this very important argument cannot be
settled at the present time. It is interesting to note from
Fig. 13 that temperatures in the outer layers of the initially
cold models, particularly one with an early crustal melting
event (model 3), eventually exceed those of the initially
molten and differentiated model 1 and thus for the last 10^9 yr
or so the "cold" initial models would have a lower viscosity in
the outer layers than the "hot" initial model.

The presence of volatiles and low melting fractions would
tend to increase the differences between the time-dependent
fractionation models and the extreme cases. The melting temperatures for all models have been assumed to be essentially
the same, but in practice this cannot be the case. The interior of a cold, undifferentiated moon would presumably have
volatile and low melting materials that, in the case of a hot,
completely differentiated moon, would have been redistributed
and concentrated in the outer layers by the differentiation
process. Thus one would expect the melting temperatures of

these models to be different. The difference cannot be determined accurately but the direction is certainly that of
increasing the melting temperature for the differentiated material and decreasing it for the undifferentiated material. This
trend would also hold within the time-dependent differentiation
models, making melting more likely before differentiation than
after.

Internally generated lava flows 3.3 to 3.8×10^9 yr ago would
not be possible for the low internal temperature models,
whereas for the initially completely molten and differentiated
model they would require an impact event or large-scale tectonic movements or both that reached to a considerable depth at
the time of occurrence of the flows on the maria surface. The
differentiating models are also consistent with the filling of
the maria surface directly after its formation as a result of
the tapping of a deep molten or partially molten layer by an
impact-generated fracture system.

These models indicate that the observed surface flows could
be produced without requiring the impacting planetismal to have
been stored in orbit for the order of a billion years after the
ending of the accretion process. The differentiating models
require only that the impact occur during the tail of the
accretion process after the surface layers had resolidified.
The fracture systems and zones of weakness could then have been
used as pathways to the surface for magma when the rising
internal temperatures increased sufficiently to melt and
mobilize the basaltic material.

Summary and Conclusions

The thermal history of the lunar interior has been investigated by many authors for many sets of parameters and initial
conditions by the construction of mathematical models. These
models have been extended to include the effects of melting and
redistribution of radioactive heat sources with time.

The models considered here include the possibility of heat
transfer by lattice conduction, radiative transfer, removal of
radioactive heat sources, and, in a molten zone, fluid convection. The energy sources are divided into two categories:
first, initial temperature sources that operate during the formation of the moon or shortly thereafter and second, long-lived
radioactive heat sources. The melting temperatures of likely
silicate materials will place an upper limit on the possible
temperature distributions within the moon. The near-surface

melting events inferred from the Apollo 11 and 12 rock samples
can provide constraints regarding the thermal history.

The two extreme initial temperature models (an initially
high-temperature, completely differentiated model, and a low-
temperature model that fails to reach melting during its
history) are unlikely on present evidence. Other representative models were constructed that provide the conditions for
surface melting at the times indicated by the experimental
evidence. These models require the movement of radioactive
heat sources within the moon with time.

Although definitive solutions to the thermal history problem
are not possible now and undoubtedly many problems will remain
with us for an indefinite time, the possibility of major
advances in our understanding in the near future is quite good.
The Apollo 11 and 12 rock samples are being subjected to
exhaustive investigations, and samples returned from other
lunar missions will presumably be available soon. Data
returned from the lunar magnetometer and seismometers are being
evaluated, and other experiments are planned including a heat-
flux measurement for the determination of the heat flow from
the interior. Such measurements, together with information
returned from other parts of the solar system, are rapidly
increasing the number of constraints any successful theory
must meet.

This increase in constraints should result in a decrease in
the number of possible models and thus permit, in the next few
years, a convergence of the interpretations of data from different sources toward a more fundamental understanding of the
origin and evolution of the moon in particular and the solar
system in general.

References

[1]Albee, A., Burnett, D. S., Chodos, A. A., Eugster, O. J.,
Huneke, J. C., Papanastassiou, D. A., Podosek, F. A., Price
Russ II, G., Sanz, H. G., Tera, F., and Wasserburg, G. W.,
"Ages, Irradiation History, and Chemical Composition of Lunar
Rocks from the Sea of Tranquility," *Science*, Vol. 167, No.
3918, Jan. 1970, pp. 463-466.

[2]Albee, A., Burnett, D. S., Chodos, A. A., Haines, E., Huneke,
J.C., Papanastassiou, D. A., Podosek, F. A., Price Russ II, G.,
Tera, F., and Wasserburg, G. W., "Rb-Sr Ages, Chemical Abun-
dance Patterns and History of Lunar Rocks," *1971 Lunar Science
Conference Abstracts*, Houston, Texas, Jan. 1971, pp. 56-57.

[3]Akimoto, S. I., Nishikawa, M., Nakamura, Y., Kushiro, I., and Katsura, T., "Melting Experiments of Lunar Crystalline Rocks," *Geochimica et Cosmochimica Acta, Suppl. 1, Proceedings of the Apollo 11 Lunar Science Conference,* 1970, pp. 129-133.

[4]Baldwin, R. B., "Summary of Arguments for a Hot Moon," *Science,* Vol. 170, No. 3964, 1970, pp. 1297-1300.

[5]Booker, J., "Thermal State of the Moon," *Transactions of the American Geophysical Union,* Vol. 51, No. 11, Nov. 1970, p. 774.

[6]Cameron, A. G. W., "Formation of the Earth-Moon System," *Transactions of the American Geophysical Union,* Vol. 51, No. 9, 1970, pp. 628-633.

[7]Clark, S. P., Jr., "Absorption Spectra of Some Silicates in the Visible and Near Infrared," *American Mineralogist,* Vol. 42, 1957, pp. 732-742.

[8]Clark, S. P., Jr., *Handbook of Physical Constants,* Memo. 97, 1966, Geological Society of America.

[9]Cohen, L. H., Ito, K., and Kennedy, G. C., "Melting and Phase Relations in an Anhydrous Basalt to 40 Kilobars," *American Journal of Science,* Vol. 265, No. 6, 1967, pp. 475-518.

[10]Compston, W., Chappell, B. W., Arriens, P. A., and Vernon, M. J., "The Chemistry and Age of Apollo 11 Lunar Material," *Geochimica et Cosmochimica Acta, Suppl. 1, Proceedings of the Apollo 11 Lunar Science Conference,* 1970, pp. 1007-1027.

[11]Darwin, G. H., "On the Secular Change in the Elements of the Orbit of a Satellite Revolving About a Tidally Distorted Planet," *Philosophical Transactions of the Royal Society,* Vol. 171, 1880, p. 713.

[12]Dyal, P. and Parkin, C. W., "The Apollo 12 Magnetometer Experiment: Internal Lunar Properties from Transient and Steady Magnetic Field Measurements," *Geochimica et Cosmochimica Acta, Suppl., Proceedings of the Apollo 12 Lunar Science Conference,* 1971 (to be published).

[13]Dyal, P., Parkin, C. W., and Sonett, C. P., "Lunar Surface Magnetometer Experiment," *Apollo 12 Preliminary Science Rept., NASA SP-235,* 1970, pp. 55-73.

[14]Fricker, P. E., Reynolds, R. T., and Summers, A. L., "On the Thermal History of the Moon," *Journal of Geophysical Research,* Vol. 72, No. 10, 1967, pp. 2649-2666.

[15]Fricker, P. E., Goldstein, J. I., and Summers, A. L., "Cooling Rates and Thermal Histories of Iron and Stony-Iron Meteorites," *Geochimica et Cosmochimica Acta,* Vol. 34, 1970, pp. 475-491.

[16]Ganapathy, R., Keays, R. R., Laub, J. C., and Anders, E., "Trace Elements in Apollo 11 Lunar Rocks: Implications for Meteorite Influx and Origin of the Moon," *Geochimica et Cosmochimica Acta, Suppl. 1, Proceedings of the Apollo 11 Lunar Science Conference,* 1970, pp. 1117-1142.

[17]Gerstenkorn, A., "Uber Gezeitenreibung beim Zweikorpenproblem," *Zeitschrift für Astrophysik,* Vol. 22, 1955, pp. 245-274.

[18]Goldreich, P., "History of the Lunar Orbit," *Reviews of Geophysics,* Vol. 4, No. 4, 1966, pp. 411-443.

[19]Gales, G. G., Randle, K., Osawa, M., Schmitt, R. A., Wakita, H., Ehmann, W. D., and Morgan, J. W., "Elemental Abundances by Instrumental Activation Analysis in Chips from 27 Lunar Rocks," *Geochimica et Cosmochimica Acta, Suppl. 1, Proceedings of the Apollo 11 Lunar Science Conference,* 1970, pp. 1165-1176.

[20]Hanks, T. C. and Anderson, D. L., "The Early Thermal History of the Earth," *Physics of the Earth and Planetary Interiors,* Vol. 2, No. 1, 1969, pp. 19-29.

[21]Hertzog, G. F. and Herman, G. F., "Na^{22}, Al^{26}, Th and U in Apollo 11 Samples," *Geochimica et Cosmochimica Acta, Suppl. 1, Proceedings of the Apollo 11 Lunar Science Conference,* 1970, pp. 1239-1245.

[22]Kanamori, H., Fujii, N., and Mizutani, H., "Thermal Diffusivity Measurement of Rock-Forming Minerals from 300° to 1100°K," *Journal of Geophysical Research,* Vol. 73, No. 2, 1968, pp. 595-605.

[23]Kaula, W. M., "Tidal Dissipation by Solid Friction and the Resulting Orbital Evolution," *Reviews of Geophysics,* Vol. 2, No. 4, 1964, pp. 661-685.

[24]Kopal, Z., "Thermal History of the Moon and of the Terrestrial Planets: Numerical Results," TR 32-225, 1962, Jet Propulsion Laboratory, California Institute of Technology.

[25]Kushiro, I., Yoder, H. S., Jr., and Nichikawa, M., "Effect of Water on the Melting of Enstatite," *Geological Society of America Bulletin,* Vol. 79, 1968, pp. 1685-1692.

[26]Levin, B. J., "Thermal History of the Moon," *The Moon,* edited by Z. Kopal and Z. K. Mikhailov, Academic Press, New York, 1962, pp. 157-167.

[27]Levin, B. J., "The Structure of the Moon," *Proceedings of the California Institute of Technology-Jet Propulsion Laboratory Lunar and Planetary Conference,* 1965, pp. 61-76.

[28]LSPET (Lunar Sample Preliminary Examination Team), "Preliminary Examination of Lunar Samples from Apollo 11," *Science,* Vol. 165, 1969, pp. 1222-1227.

[29]LSPET (Lunar Sample Preliminary Examination Team), "Preliminary Examination of Lunar Samples from Apollo 12," *Science,* Vol. 167, 1970, pp. 1325-1339.

[30]Lubimova, E. A., "Theory of Thermal State of the Earth's Mantle," *The Earth's Mantle,* edited by T. F. Gaskell, Academic Press, New York, 1967, pp. 231-323.

[31]MacDonald, G. J. F., "Calculations on the Thermal History of the Earth," *Journal of Geophysical Research,* Vol. 64, No. 11, 1959, pp. 1967-2000.

[32]MacDonald, G. J. F., "Interior of the Moon," *Science,* Vol. 133, 1961, pp. 1045-1050.

[33]MacDonald, G. J. F., "On the Internal Constitution of the Inner Planets," *Journal of Geophysical Research,* Vol. 67, No. 7, 1962, pp. 2945-2974.

[34]MacDonald, G. J. F., "Dependence of the Surface Heat Flow on the Radioactivity of the Earth," *Journal of Geophysical Research,* Vol. 69, No. 14, 1964, pp. 2933-2946.

[35]MacDonald, G. J. F., "Tidal Friction," *Reviews of Geophysics,* Vol. 2, No. 3, 1964, pp. 467-541.

[36]Maeva, S. V., "Some Calculations of the Thermal History of Mars and of the Moon," *Soviet Physics-Doklady,* Vol. 9, No. 11, 1965, pp. 945-948.

[37]McConnell, R. K., McClaine, L., Lee, D., Aaronson, J., and Allen, J., "A Model for Planetary Igneous Differentiation," *Reviews of Geophysics,* Vol. 5, No. 2, 1967, pp. 121-172.

[38]Ness, N. F., "The Electrical Conductivity and Internal Temperature of the Moon," Goddard Space Flight Center Rept. X-616-69-191 (also presented as paper K11 at XII COSPAR, Committee on Space Research, Prague, 1969).

[39]O'Keefe, J. A., "The Origin of the Moon," *Journal of Geophysical Research,* Vol. 75, No. 32, 1970, pp. 6565-6574.

[40]Opik, E. J., "The Moon's Surface," *Annual Reviews of Astronomy and Astrophysics,* Vol. 7, 1969, pp. 473-526.

[41]Phinney, R. A. and Anderson, D. L., "Internal Temperatures of the Moon," Minnesota University Rept., 1965, Tycho Meeting.

[42]Reynolds, R. T., Fricker, P. E., and Summers, A. L., "Effects of Melting upon Thermal Models of the Earth," *Journal of Geophysical Research,* Vol. 71, No. 2, 1966, pp. 573-582.

[43]Reynolds, R. T. and Summers, A. L., "Calculations on the Composition of Terrestrial Planets," *Journal of Geophysical Research,* Vol. 74, No. 10, 1969, pp. 2494-2511.

[44]Ringwood, A. E., "Petrogenesis of Apollo 11 Basalts and Implications for Lunar Origin," *Journal of Geophysical Research,* Vol. 75, No. 32, 1970, pp. 6453-6479.

[45]Ringwood, A. E., "Origin of the Moon: The Precipitation Hypothesis," *Earth and Planetary Science Letters,* Vol. 8, 1970, pp. 131-140.

[46]Ringwood, A. E. and Essene, E., "Petrogenesis of Apollo 11 Basalts, Internal Constitution and Origin of the Moon," *Geochimica et Cosmochimica Acta, Suppl. 1, Proceedings of the Apollo 11 Lunar Science Conference,* 1970, pp. 769-799.

[47]Runcorn, S. K., "Convection in the Moon," *Nature,* Vol. 195, 1962, pp. 1150-1151.

[48]Ruskol, E. L., "The Tidal History and Origin of the Earth-Moon System," *Soviet Astronomy,* Vol. 10, No. 4, 1967, pp. 659-665.

[49]Schmidt, O. U., "Genesis of Planets and Satellites," *Izvestia Academii Nauk SSSR, Ser. Fizich,* Vol. 14, 1950, pp. 29-45 (in Russian).

[50]Silver, L. T., "Uranium-Thorium-Lead Isotopes in Some Tranquility Base Samples and Their Implication for Lunar History,"

Geochimica et Cosmochimica Acta, Suppl. 1, Proceedings of the Apollo 11 Lunar Science Conference, 1970, pp. 1533-1574.

[51]Singer, S. F., "Origin of the Moon by Capture and Its Consequences," *Transactions of American Geophysical Union,* Vol. 51, No. 9, 1970, pp. 637-641.

[52]Sonett, C. P. and Colburn, D. S., "Electrical Heating of Meteorite Parent Bodies and Planets by Dynamo Induction from a Pre-Main Sequence T Tauri 'Solar Wind'," *Nature,* Vol. 219, 1968, pp. 924-926.

[53]Sonett, C. P., Dyal, P., Colburn, D. S., Smith, B. F., Schubert, G., Swartz, K., Mihalov, J. D., and Parkin, C. W., "Induced and Permanent Magnetism on the Moon: Structural and Evolutionary Implications," *Transactions of the International Astronomical Union,* 1971 (to be published).

[54]Sonett, C. P., Mihalov, J. D., and Ness, N. F., "Concerning the Electrical Conductivity of the Moon," *Journal of Geophysical Research,* 1971 (to be published).

[55]Sonett, C. P., Schubert, G., Smith, B. F., Schwartz, K., and Colburn, D. S., "Lunar Electrical Conductivity from Apollo 12 Magnetometer Measurements: Compositional and Thermal Inferences," *Geochimica et Cosmochimica Acta, Suppl., Proceedings of the Apollo 12 Lunar Science Conference,* 1971 (to be published).

[56]Tatsumoto, M., "Age of the Moon: An Isotopic Study of the U-Th-Pb Systemics of Apollo 11 Lunar Samples," *Geochimica et Cosmochimica Acta, Suppl., Proceedings of the Apollo 11 Lunar Science Conference,* 1970, pp. 1595-1612.

[57]Turcotte, D. L. and Oxburgh, E. R., "Implications of Convection Within the Moon," *Nature,* Vol. 223, 1969, pp. 250-251.

[58]Turner, G., "Argon-40/Argon-39 Dating of Lunar Rock Samples," *Geochimica et Cosmochimica Acta, Suppl. 1, Proceedings of the Apollo 11 Lunar Science Conference,* 1970, pp. 1665-1685.

[59]Turner, G., "^{40}AR-^{39}AR Ages from the Lunar Maria," *1971 Lunar Science Conference Abstracts,* Houston, Texas, Jan. 1971, p. 63.

[60]Urey, H. C., *The Planets: Their Origin and Development,* Yale University Press, New Haven, Conn., 1952.

[61]Urey, H. C., "Boundary Conditions for Theories of the Origin of the Solar System," *Physics and Chemistry of the Earth*, Vol. 2, 1957, pp. 46-76.

[62]Urey, H. C., "Origin and History of the Moon," *Physics and Astronomy of the Moon*, edited by Z. Kopal, Academic Press, New York, 1962, pp. 481-523.

[63]Urey, H. C. and MacDonald, G. J. F., "Origin and History of the Moon," *Physics and Astronomy of the Moon*, edited by Z. Kopal, 2nd ed., Academic Press, New York (to be published).

[64]Wakita, H., Schmitt, R. A., and Rey, P., "Elemental Abundances of Major, Minor and Trace Elements in Apollo 11 Lunar Rocks, Soil, and Core Samples," *Geochimica et Cosmochimica Acta, Suppl. 1, Proceedings of the Apollo 11 Lunar Science Conference*, 1970, pp. 1685-1717.

[65]Wasserburg, G. J., MacDonald, G. J. F., Hoyle, F., Fowler, W. A., "Relative Contribution of Uranium, Thorium, and Potassium to Heat Production in the Earth," *Science*, Vol. 143, 1964, pp. 465-467.

[66]Wise, D. U., "Origin of the Moon from the Earth: Some New Mechanisms and Comparisons," *Journal of Geophysical Research*, Vol. 74, No. 25, 1969, pp. 6034-6045.

[67]Wood, J. A., Dickey, J. S., Marvin, U. B., and Powell, B. N., "Lunar Anorthorites and a Geophysical Model of the Moon," *Geochimica et Cosmochimica Acta, Suppl. 1, Proceedings of the Apollo 11 Lunar Science Conference*, 1970.

[68]Wood, J. A., "Thermal History and Early Magnetism in the Moon," *The Geophysical Interpretation of the Moon*, edited by G. Simmons, preprint Sept. 1970 (to be published).

[69]Wyllie, P. J. and Tuttle, O. F., "Experimental Investigation of Silicate Systems Containing Two Volatile Components. Pt. 1: Geometric Considerations," *American Journal of Science*, Vol. 258, 1960, pp. 498-517.

[70]Yoder, H. S., Jr. and Tilley, C. E., "Origin of Basalt Magmas, and Experimental Study of Natural and Synthetic Rock Systems," *Journal of Petrology*, Vol. 3, No. 3, 1962, pp. 342-532.

Index to
Contributors to Volume 28

Stimpson, Leonard D.
JET PROPULSION LABORATORY
121

Summers, Audrey L.
NASA AMES RESEARCH CENTER
303

Thompson, Thomas W.
JET PROPULSION LABORATORY
83

Wechsler, Alfred E.
ARTHUR D. LITTLE, INC.
215

Winter, Donald F.
UNIVERSITY OF WASHINGTON
269

Zisk, Sidney H.
M.I.T. HAYSTACK OBSERVATORY
83